SOLUTIONS FOR CARBON REDUCTION

SOLUTIONS FOR CARBON REDUCTION

Edited by

Dr. Kailash Malode

Delve
PUBLISHING

www.delvepublishing.com

Solutions for Carbon Reduction

Dr. Kailash Malode

Delve Publishing
224 Shoreacres Road
Burlington, ON L7L 2H2
Canada
www.delvepublishing.com
Email: orders@arclereducation.com

ISBN: 978-1-77469-903-4 (Hardcover)

ABOUT THE EDITOR

Dr. Kailash Malode (1980) is presently serving as Assistant Professor, Department of Soil Science And Agriculture Chemistry, college of Agriculture, Parul University, Vadodara, Gujarat. He obtained his B. Sc.(Ag.) in 2006 from Marathwada Agriculture University, Parbhani. M. Sc. (Ag.)2009 in Dr. Panjabrao Deshmukh Krishi Vidyapeeth, Akola and Ph. D. Soil Science And Agriculture Chemistry in 2014 from Vasantrao Naik Marathwada Krishi Vidhyapeeth, Parbhani. He started his career as Assistant professor in 2014. His field of specialization is soil fertility, soil carbon sequestration and organic carbon management. He has edited 06 books at international level, published 19 research papers in National and international journals of reputed. He has participated in many National, State Seminars and symposiums. He has written 02 practical manuals for the undergraduate course for the benefit of students, besides several radio talks and on farmers on a farmer training programme. Dr. Kailash has also written 26 popular articles in Shati Bharti, Shati pragati and other reputed magazines and 22 audio talks. He had the Best writer award in 2015 for Krishi dooth popular Magazines of Maharashtra, Young Scientists Award 2021 by Astha foundation and Birsa Agriculture University, Jharkhand. Dr. Kailash got the Best scientist award in soil science 2023 by Rajmata Vijayaraje Sindhya Krishi Vishva Vidhyalaya, Gwalior, and also best poster presentation of same.

TABLE OF CONTENTS

LIST OF FIGURES

PREFACE

The Earth is being subjected to global warming as a result of CO_2 emissions from fossil fuels industry. To comply with the Paris Agreement and the UN's Sustainable Development Goals (SDGs), it is necessary to reduce these CO_2 emissions as well as those of other greenhouse gases such as methane, NOX, and CFCs. Even though some technologies are highly advanced, sustainable carbon dioxide capture remains difficult. The inspiration for this book comes from recent advancements that point to a new set of technologies that can make carbon capture both environmentally and financially practical. This book fills in a knowledge gap between the contribution of novel materials and cutting-edge technologies to sustainability and offers the most recent information on breakthroughs in carbon capture for mitigating climate change. It also covers recently discovered uses for carbon capture in the production of hydrogen, biofuels, and carbon-neutral fuels. The detrimental effects of human-caused greenhouse gas (GHG) emissions are fundamentally altering the social, political, and natural landscape. Numerous studies have connected these emissions to environmental damage, animal extinction, extreme weather patterns, and global warming. Governments all across the globe have created laws to track and cut carbon emissions, most notably the Kyoto Treaty, to lessen these adverse effects. Climate change is one of our generation's most urgent sustainability issues, and this is understood by many governmental, business, nonprofit, and local groups. They have been keeping track of their GHG emissions and have started to make preparations for a world marked by energy and climate issues. The significance and necessity of accurately measuring and reducing energy and GHG emissions are reiterated by these advancements.

The pertinent information and skill set for managing individual carbon footprints have been published in fragments in a variety of papers, reports, and guidelines, despite the fact that the matter is urgent, timely, and generally practiced in a disorganized manner. It is challenging for a person or an organization to comprehend how the many types of methods work together to reduce their carbon footprint. This book will help students who are interested in studying climate change processes and willing to lower the menace of carbon dioxide emissions that is destroying the natural resources that humans rely upon.

Introduction

There are a lot of possibilities if we were to compare the efforts and opportunities for CO_2 capture. To reduce CO_2 emissions, however, we must also take into account the process requirements, resources, costs, and life cycle effectiveness. On the path forward, several materials can be simulated, blended, and synthesized to acquire the optimum practice for each requirement. Therefore, it is always beneficial to create new materials by experimenting with cutting-edge techniques and combinations, with CO_2 capture with a sustainable perspective remaining everyone's top priority. We must look at additional cross-disciplinary and geographically diverse connections between various strategies for carbon capture and the implementation of Sustainable Development Goals (SDGs). In order to promote global climate control, this may also entail creating shared evaluation standards or metrics to gauge the environmental effects of sustainable development policies. One percent of total emissions were due to the usage of paper. The objective of minimizing paper usage has been made simpler by advances in computer technology during the past 20 years. Reducing the amount of paper purchased annually and encouraging the university community to print fewer pages are two reduction techniques for this category. One metric tonne of CO2e is represented by one offset credit. Since it is practically impossible for a university/company/organization to completely stop producing GHGs, buying carbon offsets will be essential to their effort to become carbon neutral. The systems approach, which forms the foundation of the solution approaches, is a key idea in the successful implementation of energy and carbon footprint minimization. This strategy addresses concerns with emission rates and energy use at the earliest stages feasible while taking into account the whole supply chain. Numerous parallel initiatives have been made through national and local policy acts, the corporate sector, as well as the United Nations Framework Convention on Climate Change (UNFCCC) to collate the demand for GHG mitigation. Both the public and private sectors are now developing standards that could serve as the cornerstone for the broad adoption of emissions trading as a method for efficiently reducing GHG emissions. Legal trading instruments are being established, some markets are being created all over the world, and derivative markets are already being examined. Forest ecosystems are particularly well known for containing a sizeable portion of the global biodiversity, which is essential for preserving ecosystem balance, evolutionary processes, and natural shock resistance. At the local level, forests are credited with sustaining a variety of watershed services, including flood protection, stream flow regulation, soil erosion control, and aquatic productivity maintenance.

CHAPTER 1

Climate Change

CONTENTS

1.1 INTRODUCTION

Most people agree that climate change is the most significant environmental issue for humans. International collaboration to reduce greenhouse gases is crucial since the atmosphere recognizes no borders, and trade and money movements connect the globaleconomies (Rosswall, 2002; Oberthür, 2003). The 1992 United Nations Foundation Convention on Climate Change (UNFCCC), which entered into force ten years ago(Seo, 2017), and its addendum, the 1997 Kyoto Protocol, offer the institutional framework for such cooperation(Rosen, 2015)efficiency, and effectiveness, I analyze the Kyoto Protocol and argue it is a clear case of institutional failure, with the design itself bearing substantial blame for this outcome. The study points to how particular features of the Protocol—its short time frame for action, binding targets, emission reduction measures, and provision for future commitment periods—have resulted in short-sighted behavior by member states and path-dependent structures that failed to make a substantial impact on the climate problem. Related Articles Viniegra, Ibarrarán, María Eugenia, and Salimah Mónica Cossens González. 2007. Climate Change Research and Policy in Mexico: Implications for North American Security. Politics & Policy 35 (4. The Kyoto targets can be met through domestic efforts in the countries that have them. By utilizing the three adaptable methods outlined in the protocol, joint implementation (JI), the Clean Development Mechanism (CDM), and international emissions trading (IET), they can also reduce greenhouse gases (GHG) (Figure 1.1) in other nations at a cheaper cost than at home (Youngman, Schmidt, Lee, & De Coninck, 2007).

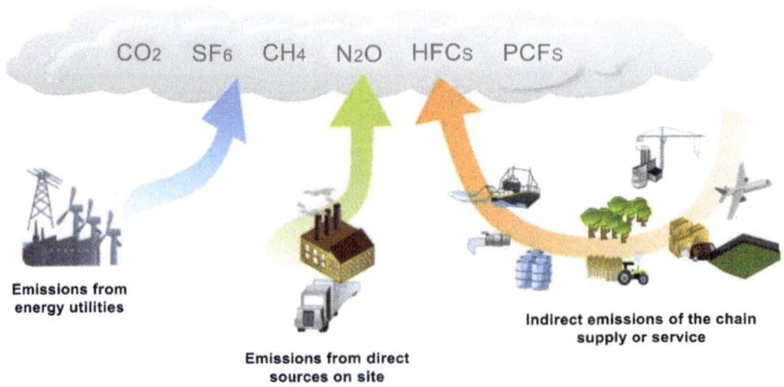

Figure 1.1. Greenhouse gases. (Source: Dadkhah M, Tulliani J-M. Nanostructured Metal Oxide Semiconductors towards Greenhouse Gas Detection. *Chemosensors*. 2022; 10(2):57. https://doi.org/10.3390/chemosensors10020057, Creative Commons License)

Cost-effectiveness arguments favor allowing governments to claim credit for overseas efforts that reduce GHG emissions at source or improve the removal of GHGs by sinks because the location of GHG emissions is environmentally immaterial. Although there are also political, economic, social, and ethical reasons why domestic action should be preferred, one of the most striking features of the protocol is how its flexible mechanisms are being implemented quickly worldwide, even before it officially enters into force.

The Conference of the Parties (COP) (UNFCCC), which serves as the interim institutional body of the protocol, has legal competence to execute the Clean Development Mechanism (CDM) under the Kyoto Protocol, one of its mechanisms, in more than 50 developing nations(Churchill & Ulfstein, 2000). A plan for trading GHG emission allowances within the 25 Member States of the European Union (EU), which began on 1 January 2005, is even more significant An estimated 15,000 firms are covered by the EU Emission Allowance Trading Scheme (EU ETS), which has a market potential of 10 billion euros annually. The EU program connects to the developing JI market centered on Central and Eastern European nations and the current CDM market. In addition, domestic trading and offset schemes are being developed in other regions of the world, such as the US and Australia, that are Kyoto compliant and, in many cases, anticipate future linkages with the Kyoto mechanisms, despite the Bush administration's rejection of the protocol.

1.2 REGULATIONS

These regulations cover three distinct legal spheres: national schemes established at the federal or state level that establish trading or offset provisions for national entities; regional regulations, primarily those agreed upon by the European Union governing the EU ETS; and international regulations as outlined in the Kyoto Protocol and elaborated through decisions made by the COP of the UNFCCC (E. Boyd, Corbera, & Estrada, 2008).These programs' carbon markets will interact in intricate ways. Cost-reduction will be the main driving force behind exchanges among and between schemes, especially at the entity level, but legal and institutional provisions, as well as more general political and strategic considerations, will also play a significant factor in whether, where, when, and how trades take place. One of the advantages of this literature is that it combines advances related to the Kyoto mechanisms from the national, EU, and international levels in one location, making it easier to grasp the connections between these levels. By doing this, it hopes to assist the expanding group of policymakers in wealthy and developing nations in comprehending the regulations controlling new carbon markets and how they will operate and develop in the future. It will be helpful for businesses, market intermediaries, and service providers, including brokers, traders, auditing and certification entities, consultants, law offices, and stakeholder organizations already operating under the Kyoto protocol.

1.3 NEGOTIATIONS

Negotiations on the mechanisms turned out to be contentious in the lead-up to Kyoto for several reasons, as mentioned below. Before Annex I Parties felt secure enough to start the domestic ratification process for the protocol, the mechanisms-related language provisions in Articles 6, 12, and 17 of the Protocol needed to be substantially developed (Littlejohns et al., 2020). It took slightly under six years to develop the international rules and protocols for the extra mechanisms. With the approval of the Bonn Agreements at COP-6 Part II in July 2001, most critical mechanisms-related issues were eventually settled.

Detailed textual choices were outlined in the Marrakesh Accords, which were accepted at COP-7 (Huq, 2002). During COP-9, which took place in Milan in December 2003, issues regarding the inclusion of sinks in the CDM were settled. Many of the international regulations of the protocol are proposed draft decisions that are adopted by the COP and are then sent by the COP to the COP/MOP for formal adoption at their first meeting. Since they are a part of the carefully considered package of decisions adopted in accordance with the Bonn Agreements and the Marrakesh Accords, the COP decisions containing the recommended decisions COP/MOP-1 are anticipated to be confirmed in full by the COP/MOP without further discussion, despite not yet having full legal force. As a result, there are now two distinct treaties, as well as more than a dozen COP resolutions totaling more than 200 pages of text, that contain the rules, institutions, and procedures for managing access to and use of the Kyoto flexibility mechanisms (Firger & Gerrard, 2010). The Executive Board of the CDM (EB/CDM) further defines the rules governing the registration and validation of CDM projects. International mechanisms-related norms are hard to find because of these dispersed, intricately connected outputs and processes. Since these projects are the indications of the Kyoto mechanisms and many of them could be transformed into JI or CDM projects, Countries' approaches to the talks on the Kyoto mechanisms have been impacted by their domestic experience with market-based tools and carbon trading as well as fundamental disparities in regulatory culture and administrative capacity. In general, in the late 1980s and early 1990s, many non-EU nations, especially the US, had positive experiences employing market-based instruments in the context of the environment. These nations were already replacing or enhancing old command and control methods in their environmental policies with market-based tools like trading, taxes, and levies. As a result, many policy scholars, industry analysts, and regulators now view command and control legislation as disproportionately technology-prescriptive and unwelcoming to innovation.

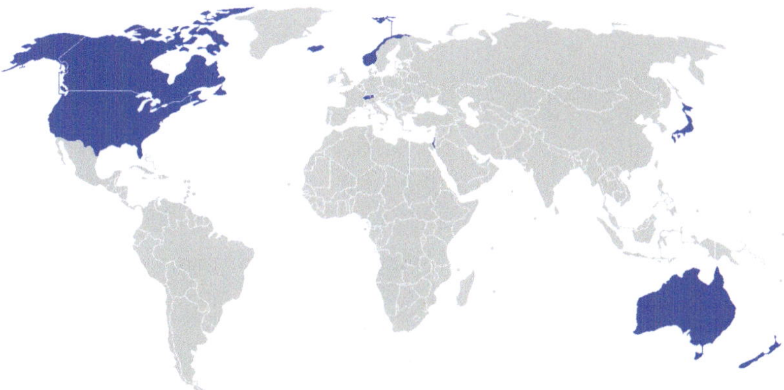

Figure 1.2. JUSCANZ group of nations. (Source: https://upload.wikimedia.org/wikipedia/commons/1/18/JUSCANZ_Countries.png, Creative Commons License)

The US and its leadership in the JUSCANZ group (originally consisted of Japan, United States, Canada, Australia and New Zealand, hence, the acronym) of nations (Figure 1.2) was crucial in persuading others of the benefits of adopting mechanisms to tackle climate change(Yamin, 2000). In the context of the 1990 modifications to the Clean Air Act, the US has previously established legal frameworks for two types of trading programs. This legislation established a domestic "cap-and-trade" system that involved limiting overall emissions, allocating permits proportionately to allowable emissions, and requiring entities to possess enough permits to cover their emissions during a specified compliance period. It also established a "baseline and credit" system in which participants received "credits" for emissions reductions made in comparison to an arbitrary baseline. The production by governments of a financial asset known as a permit or allowance, which is typically given (rather than auctioned) to polluters in a cap-and-trade scheme or issued in the case of baseline and credit schemes upon proof that emissions have been reduced, is a feature shared by both programs. The Clinton Administration's experience with the substandard implementation of a carbon/energy tax in the early 1990s (later known as the British thermal unit (BTU) tax) persuaded US policymakers that any domestic US program to decrease GHGs would require emissions trading. Emissions trading would be a practical approach to reduce the costs of worldwide compliance, mainly because the US contributes almost 20% of the world's total GHG emissions (Schmalensee et al, 1998).

1.4 UNFCCC

Even while several scholars were praising the US sulfur trading program as a ground-breaking example for climate negotiators throughout the UNFCCC discussions, these early ideas were not sufficiently understood for Parties to take into account their worldwide application in the UNFCCC (Figure 1.5).

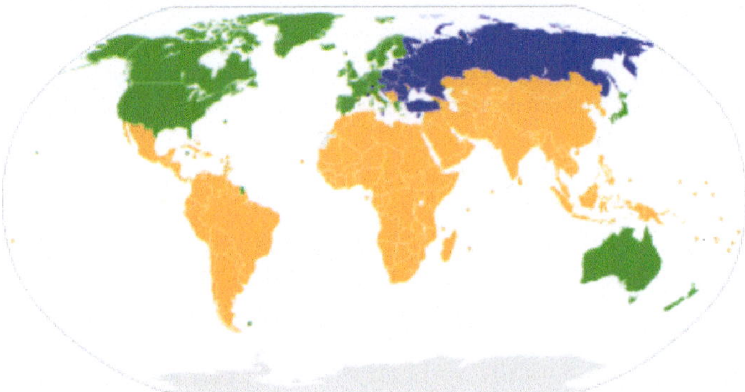

Parties to the UNFCCC

Annex I and II parties

Annex I parties

Non-annex parties

Observer states

Figure 1.5. UNFCCC. (Source: By BlankMap-World6.svg: Canuckguy (talk) and many others (see File history)derivative work: Danlaycock - This file was derived from: BlankMap-World6.svg:, Public Domain, https://commons.wikimedia.org/w/index.php?curid=31436070, Creative Commons License)

The Convention did, however, make mention of permitting groups of Annex I Parties to work together with other Parties to meet their goal of reducing their emissions of CO_2 and other GHGs to 1990 levels by 2000. However, the Convention's use of the term "joint implementation" in Article 4.2(a) and 4.2(d) was unclear because it might be interpreted to mean merely cap-and-trade and baseline and credit forms of international trading in addition to a wide range of other forms of cooperation(Smith, 2009). Because of this, it was challenging to operationalize the requirements for joint implementation during the 1995 COP-1 conference in Berlin. Even though some developing nations, like Costa Rica and other nations in Central and Latin America, had by that point come out in favor of project-based forms of trade, the majority of developing nations were opposed to them because Annex I Parties had not at Berlin committed themselves to binding legal targets. In addition, the introduction of licenses and credits on an international scale prompted questions about how to distribute them among Parties with very varied legal obligations as well as fundamental ethical, equitable, and environmental concerns. Developing countries were skeptical that these wealthy countries would use trade to avoid taking domestic action while shifting the actual burden of pollution control to the South because JUSCANNZ was also pushing hard for developing countries to take on additional commitments and was the leading advocate for mechanisms.

Following COP-1, the Ad Hoc Group on the Berlin Mandate started negotiating the obligations of the Annex I Parties(Matsuo, 2003). JUSCANZ persisted in advocating for project-based systems and emissions trading, as well as sometimes calling for commitments from developing nations. The hostile and distrustful environment that resulted highlighted the skepticism that poor countries had over the significance that JUSCANZ gave to the procedures. The regulatory culture of the EU Member States as a whole was less welcoming of trading instruments, despite the fact that certain nations were moving toward the use of environmental taxes and other forms of charge, even though differences of opinion started to emerge within the EU. On the other hand, developing nations continued to be suspicious or hostile for the reasons mentioned above. Additionally, the majority of developing nations were having difficulty implementing first-generation environmental legislation to address severe environmental issues including water pollution and local air quality when AGBM negotiations started nearly ten years ago. As a result, they had little time to learn more about cutting-edge methods for mitigating the effects of climate change, a problem they believed the North should tackle first and one for which the use of experimental approaches with their limited regulatory resources did not make sense given its long-term nature. The academic and policy literature on the mechanisms did not help because much of it was vehemently pro-taxes or pro-emissions trading with little discussion of how different instruments could be mixed, matched, and overlaid in the context of existing regulations to suit countries' extremely diverse legal, administrative, and regulatory structures. The US announced at COP-2 in 1996 that it sought to negotiate an instrument with legally binding targets and some type of carbon trading that served as a signal that the protocol's negotiations would need to take emissions trading seriously(Bodansky, 2016).

A US document that was presented in December 1996 in response to this one formally tied this viewpoint to the introduction of carbon pricing. As the EU continued to promote obligatory policies and measures (PAMs) while adding its voice to developing nations' concerns that domestic action should be prioritized, the US position shifted polarized differences within the OECD. As talks carried on into 1997, it became evident that several EITs would be seeking targets that would result in their receiving significant allowances that they could exchange with others. The Umbrella group, which consists of JUSCANZ, Russia, and Ukraine, was formed, raising severe concerns about the negative impact of air on the environmental effectiveness of the new regime.

The argument that wealthy nations would exploit the mechanisms to avoid enacting politically unpleasant domestic reductions fueled criticism from the EU and underdeveloped nations. Questions about whether domestic trading programs, like the sulfur program, could be transferred to the global context in the somewhat simplistic ways that some economists recommended also started to be questioned by the scientific community. This is due to the fact that monitoring, tracking, and verifying GHG emissions in an international setting is far more difficult than it is in local trading schemes, which rely on severe fines and tight enforcement to dissuade non-compliance. In the absence of an international body mandating compliance, accurate and timely self-reporting by governments has a poor track record. These uncertainties were made worse by the fact that many Annex I Parties had fallen behind on their reporting obligations under the Convention, giving rise to doubts that trading mechanisms that required governments to monitor thousands of emissions sources would prove to be too demanding for many governments and might give rise to fraudulent transactions, undermining the environmental integrity of the regime.

1.5 SITUATION IN DEVELOPING COUNTRIES

Even though developing nations continued to oppose cap-and-trade forms of emissions trading and in fact did not take part in the Kyoto Protocol drafting groups dealing with Joint Implementation (JI) and International Emissions Trading (IET), many of them began to show an increasing amount of interest in project-based mechanisms, which was encouraged by the growth of AIJ projects that aided institutional learning(Atle C. Christiansen & Wettestad, 2003). Brazil took the initiative in outlining the key components of the CDM during the final six months of 1997. The CDM is especially well suited to advancing quantified reductions of GHG in large, industrializing developing countries with rapidly expanding economies that are inherently difficult to manage due to the high levels of domestic and foreign capital investments they involve. This is evidenced by the financial crisis, which devastated many economies in Asia and Latin America.

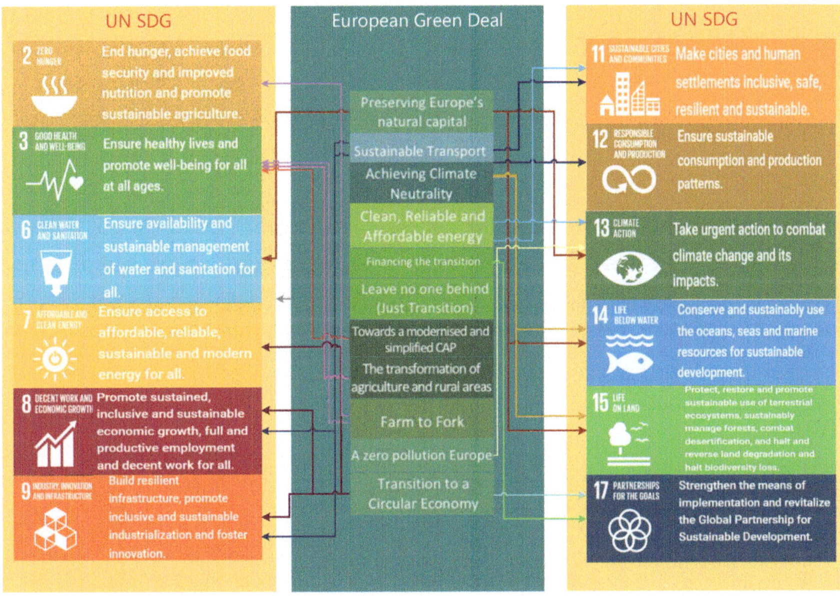

Figure 1.6. EU and sustainable development goals (SDGs). (Source: Ionescu RV, Antohi VM, Zlati ML, Georgescu LP, Iticescu C. To a Green Economy across the European Union

Although the underlying opposition of developing nations toward trade nearly prevented the inclusion of Article 17 from being agreed upon, COP-3 ended with the addition of the three mechanisms and an explanation of how the EU's joint fulfillment would operate in practice in Article 4(Gupta,

2010). The conclusion of the Kyoto negotiations marked a significant turning point in the mechanisms discussions. As groups of Parties started to assemble around their preferred mechanism, the answers to this question changed. The CDM was naturally insisted upon as a top priority by poor countries due to their limited negotiating resources and economic interests. JUSCANNZ continued to emphasize carbon trading while dismissing project-based mechanisms as a subpar type of policy due to their greater transaction costs(Wettestad, 2009). The smaller EITs worked hard to prevent JI from being squeezed by IET and CDM as these two mechanisms began to demand the lion's share of policy attention and negotiating time, aware of the challenges of assembling credible inventory data that would allow them to participate in IET.

1.6 EU

The EU (Figure 1.6) embarked on attempting to make everyone happy while trying to maintain its environmental credentials during this time, even though the bargaining stance of JUSCANZ and poor countries was quite apparent. Even though environmental NGOs supported its position that domestic action should come first, it was difficult to reconcile that position with the interests of its negotiating partners and, over time, with the growing domestic interest in emissions trading at both the Member State and Commission levels. With the implementation of the European Union Emissions Trading System (EU ETS), many developing nations were concerned that the CDM would not be forced out of the market as it would be if the proposed across-the-board limits of the EU on the mechanisms were to be implemented. The EITs, that had begun accession talks with the EU, were also concerned that the market for Joint implementation might end up being quite tiny. An intense period of learning was evident among negotiators, GHG emitting industries, stakeholders, and a variety of market intermediaries who all realized that the Kyoto mechanisms could be good news for business, or at least better than the other alternatives. At the same time, the international agenda was clogged with numerous conferences, workshops, and intersessional meetings discussing mechanism design issues. The Kyoto mechanisms, which were gaining international support, experienced two significant defeats around this time. The first was the decision made at COP-6 Part II and COP-7 on the inclusion of sinks under Articles 3.3 and 3.4 of the Protocol. The availability of sinks for the majority of Parties is the main element affecting compliance costs, and this has a crucial consequence on the extent to which each country needs to take additional GHG reduction action, whether domestically or abroad. The final agreement reached in the Bonn Agreement and Marrakesh Accords indicates Annex I Parties will have less need to employ the mechanisms; instead, many of them are likely to first consider low- or no-cost sinks choices unless there is domestic pressure compelling them otherwise(Zhang & Baranzini, 2004). This pool of inexpensive CERs is made much larger by the inclusion of sinks in the CDM, which was agreed upon in principle at COP-6 and in detail at COP-9. The Bush Administration's declaration that it will not ratify the Protocol in March 2001 was the second setback. The US declaration, as the largest GHG producer and possibly the biggest buyer of Kyoto units, sent the climate system into meltdown and reduced hopes for the feasibility and scale of the developing carbon markets.

However, the protocol survived the Bush Administration's declaration that it was flawed. Instead, it had the exact opposite effect, inspiring the rest of the world to defend the protocol, a document that had taken over a decade to negotiate because of its multilateralism and genuine care for the environment worldwide. The Bush Administration's inability to develop a politically viable alternative in time for COP-6 Part II in July 2001 was quickly overshadowed by the Administration's focus on responding to the events of September 11th, which occurred a few weeks before COP-7, which was held in Marrakesh(Cass, 2005). For the entire climate community, the EU's leadership on the procedures starting in 2001 was a huge plus. Interest in Kyoto and its methods continued to be high after the EU announced that it will launch a domestic carbon trading program starting in 2005, covering at least 25 countries. The World Bank's creation of the Prototype Carbon Fund (PCF) and the measures taken by numerous governments to conduct tenders for project-based credits, such as the Dutch ERUPT and CERUPT programs, also indicated widespread support for the continuance of the Kyoto mechanisms(Michaelowa, 2003). The Marrakesh Accords, which were adopted by COP-7 and contain almost 100 pages of text operationalizing all three mechanisms and complex interlinkages with the crucial compliance issues, caused the focus to shift once more to a higher level of attention as technical concerns about registry design and baseline methodologies now preoccupied the mechanisms community.

The smooth start of the CDM Executive Board(CDM EB)shows a clear desire on the part of the Parties to quickly activate the Kyoto mechanisms. It has done so despite the interferences brought on by the Bush Administration's insistence that the US not provide financial support for CDM activities, a stance that is difficult to understand given that the CDM directly contributes to GHG reductions by the developing nations that the Administration criticized for not doing enough when the US backtracked on Kyoto. More and more evidence was emerging that the Bush Administration's disruptive strategies and non-engagement with the climate crisis are at odds not only with scientific opinion and the international community but also with the general direction of climate policy in the rest of the US. Over 45 bills, resolutions, and amendments addressing climate change were introduced in the US Congress in 2003–2004(Fisher, Leifeld, & Iwaki, 2013).There are already many state initiatives in place in the US that show interest in the Kyoto processes. Implementation efforts are also advancing at the domestic level in other countries as they start the practical task of designating national authorities and equipping them with resources and legal authority to carry out the functions outlined for them in the Marrakesh Accords. Despite significant policy uncertainties, the private sector as a whole has contributed positively. Many major players have faced high learning and transaction costs in order to put together risky CDM projects, even though they are fully aware that these will be closely scrutinized by both regulators and environmental NGOs. With the adoption of a proposal to connect the EU ETS with the Kyoto mechanisms that was put up in late 2003, the EU's leadership on the mechanisms has persisted. Despite the fact that this will strengthen the JI and CDM markets, the connection has also highlighted the challenges the EU faces in advancing the operation of the Kyoto markets prior to the protocol's entry into force. Even a program as large as the EU ETS is unable to stifle the potential supply of Kyoto units from non-EU countries(Sterk & Schüle, 2009).

1.7 CDM

The planned exclusion of sinks projects from the EU ETS under the Linking Directive proposal, however sensible from an environmental perspective, may lower expectations for several pro-CDM sinks countries about the inclusion of sinks in the CDM as agreed by COP-9. Japan and other non-EU pro-Kyoto nations have expressed concerns that they may be denied access to JI credits as a result of the operation of the EU ETS. For instance, there are underlying political issues regarding how prominently the EU scheme (a domestic scheme) should be featured. Once Kyoto comes into effect and JI and IET take center stage, these tensions are likely to decrease. Looking ahead, it is evident that if policymakers fail to provide a market value for emissions reductions in the post-2012 period, interest in the Kyoto project-based mechanisms, particularly the CDM, would wane over the next few years(Korppoo & Gassan-Zade, 2014)the role of the private sector in their implementation, the coordination required, and the political will gained. Even though a weak rule of law, problems with policy implementation, and the ambiguous role of private-sector actors are not 'make or break' issues, they are likely to defer future mechanisms. Success and failure will, rather, hinge on the priority these factors are accorded by the top leadership. It is likely that simple mechanisms that only involve a few actors will be less complicated to set up and run than, for instance, emissions trading schemes (which require domestic burden sharing. Investors won't put much money into the CDM until it is guaranteed that CERs will be valuable and can be used as cash to offset future targets. In order to address the post-2012 mitigation pledges of Annex I Parties, international agreements are necessary. The year 2012 broughtup the issue of US involvement and provide developing nations the chance to influence the development of the Kyoto mechanisms and their role in the creation of the future climate regime, many nations are reluctant to discuss it. According to Article 4.2(a) and (b) of the Convention, Annex I Parties must adopt policies and actions to reduce their GHG emissions and to change the long-term pattern of their emissions by putting PAMs into practice(Metz et al, 2002)capability and development needs. Each of these, separately or in combination, has been used in designing schemes for differentiation of commitments to limit or reduce greenhouse gas emissions. Many implementation problems of these proposals are often side-stepped. In particular, some proposals may be incompatible with Article 2 of the United Nations Framework Convention on Climate Change (UNFCCC (Metz, Berk, den Elzen, de Vries, & van Vuuren, 2002). This includes accomplishing the quantified target of the Convention, which is to return CO_2 and other GHG to 1990 levels by the year 2000(Breidenich et al, 1998) (Breidenich, Magraw, Rowley, & Rubin, 1998).Such Parties may implement such policies and measures jointly with other Parties, according to Article 4.2(a), however, this clause is subject to whatever criteria the COP may establish in accordance with Article 4.2(d) respecting joint implementation. Without approval from the COP, different governments, companies, and NGOs can work together to decrease or sequester GHGs. Although different interpretations are possible, one goal of Article 4.2(d) was to give Parties a generally recognized means to receive credit for implementing or financing joint actions. Identification of which activities should be considered by Annex II Parties as fulfilling their mandatory Convention commitments and which should earn credits proved problematic and continues to prove challenging in the context of consideration of how the issues of additionality of AIJ projects should be assessed. This is because the Convention requires Annex II Parties to provide funding and technological assistance to developing country Parties to help the latter meet their obligations. On the fundamental question of when Annex I Parties should request credits that may be used to fulfill their UNFCCC mitigation commitments, COP-1 was unable to come to a consensus. The granting of credits was rejected by many developing nations as well, thus at COP-1 it was decided that no credits shall accrue to any Party as a result of AIJ operations during the pilot phase. This phase was supposed to go through a thorough evaluation and was supposed to be finished by the end of 1999. The timing was chosen so that the pilot period could provide enough "learning by doing" and that protocol talks addressing political difficulties over the overall size and burden-sharing agreements relating to Annex I Party emissions targets could be concluded.

As was mentioned above, several poor nations and the EU were worried about making sure that their domestic and international mitigation abatement activities were balanced appropriately. Since COP-1, the COP has periodically evaluated the geographic distribution of projects as well as the manner in which information on AIJ initiatives should be supplied. The pilot phase was examined by the COP-5, which was held in 1999, and it was resolved to continue it after 1999, urging Parties that had not yet participated in projects to do so(Tangen, 1999). The COP-7 and COP-8 have likewise decided to continue the pilot phase. Prior to COP-8, Parties taking part in AIJ projects were asked to supply the Secretariat with AIJ data so that it could create an annual document that was taken into consideration by each COP. The COP-8 determined that going forward, the synthesis paper shall be prepared on a biennial basis rather than an annual basis as originally specified in Decision 5/CP.1. This will simplify and decrease such reporting. In Decision 5/CP.1, the UNFCCC Subsidiary Bodies (SBs) were requested to create a framework for Parties to report on "the potential global benefits and the national economic, social, and environmental impacts, as well as any practical experience gained or technical difficulties encountered," in a transparent and reliable manner. Such reporting was essential in deciding how baselines that satisfied the 'additionality' standards for finances and the environment stipulated by COP-1 were to be evaluated. Such reporting was and continues to be separate from information supplied in the national communications of all Parties because participation in AIJ is voluntary. The first universal reporting format (URF), which Parties might use to voluntarily report on their AIJ projects, was adopted at COP-3. The COP-8 adopted a revised URF that integrates lessons learned from the initial URF and takes into account the type of data that would be required in the future to determine whether AIJ projects can achieve CDM and JI eligibility requirements(Grainger, 2015)to be realized by reducing the rate of land degradation and increasing the rate of restoration of degraded land, was agreed at the Rio+20 Conference in 2012. This paper evaluates the feasibility of introducing a Land Degradation Neutral (LDN.

Despite the fact that COP-1 was unable to agree on credits, Decision 5/CP.1 set new standards for AIJ, many of which led into or were immediately included in CDM and JI modalities. Negotiations on reporting and methodological concerns connected to the CDM and JI projects were aided by experience building a URF for reporting AIJ projects. The actual requirements for starting AIJ initiatives are still as stated in Decision 5/CP.1 and include the following:AIJ can take place between Annex I Parties and with non-Annex I Parties who desire it; AIJ projects are supplemental to and just one subsidiary means of attaining the Convention's purpose; and AIJ in no way changes the commitments of any Party under the Convention. Participation in AIJ is voluntary for all Parties and requires prior acceptance, approval, or endorsement by the governments of Parties concerned. The financial additionality requirement, according to which funding of AIJ shall be additional to the financial responsibilities of [Annex II Parties] within the scope of the financial mechanism as well as to present official development assistance (ODA) flows, should be met by AIJ projects. The JI and CDM modalities, which are described in more detail below, now include the environmental and financial additionality criteria for AIJ projects. The scope of AIJ is not constrained by Decision 5/CP.1. As a result, in theory, any project that reduces or sequesters emissions can qualify as an AIJ project. A significant portion of the approximately 150 AIJ initiatives that have been formally reported to the UNFCCC Secretariat are in EITs(Stowell, 2005). The number, location, and project type of AIJ projects are routinely updated and reported in JI Quarterly. Although AIJ is not specifically mentioned in the CDM and JI modalities and procedures, COP-7 agreed on the following. To qualify for validation and registration as a clean development mechanism project, a project activity must have begun as of the year 2000 and before the implementation of this decision. It must also have been submitted for registration by the 31st of December 2005. If such project activities are registered, the crediting period may begin prior to the date of registration, but no sooner than January 1, 2000.

1.8 COP

The COP-9 decided that these CDM project activities may use a crediting period beginning before the date of registration if the project activity is submitted for registration before 31 December 2005. The COP-9 explained that this expression unintentionally excluded the possibility of projects starting between the date of adoption of Decision 17/CP.7 and the date of the first registration of a CDM project activity earning CERs. If JI initiatives begin after the year 2000 and match the criteria outlined in the COP-7-agreed implementation guidelines for Article 6 of the Kyoto Protocol, they may qualify as Article 6 projects. As there is no specified amount prior to 2008 from which they can be issued, ERUs may only be granted for a crediting period commencing after that date. Because of these rules, AIJ projects cannot automatically become CDM or JI projects, as some Parties had hoped(Woerdman, 2001)transaction costs (e.g. to find a trading partner. Instead, each project must meet the requirements for CDM and JI projects as established by the COP-7 agreement. The requirements of Article 12.10 of the Protocol, which stipulate that CERs obtained during the period of 2000 to 2008 can be used by Annex I Parties for compliance with Article 3 obligations, are operationalized by the crediting start date of 2000 for CDM activities. According to Article 6, there is no comparable start date for JI activities. A Party will not have an allotted amount before 2008, hence AIJ projects conducted between Annex I Parties that qualify as JI projects will only generate ERUs after 2008. There are several common traits among the Kyoto mechanisms. Numerous of them are outlined in Decision 15/CP.7, which covers the fundamentals, range, and makeup of all three mechanisms. The Decision 19/CP.7 accounting of assigned amounts regulations, the COP-8 registry rules, and the information-related reporting and review obligations under Articles 5, 7, and 8 of the Protocol are some additional rules with a greater technical component.

The modalities for all three mechanisms include essentially identical review provisions that state: any future revisions of the modalities, rules, and procedures for each mechanism shall be decided in accordance with the COP/MOP as applied; the first such reSview will be carried out no later than one year following the conclusion of the first commitment period (i.e., 2013), based on recommendations by the SBI, drawing on technical advice from SBSTA. Any adjustments brought about by overall reviews will not affect projects that have already started, according to the JI and CDM protocols (Figure 1.7).

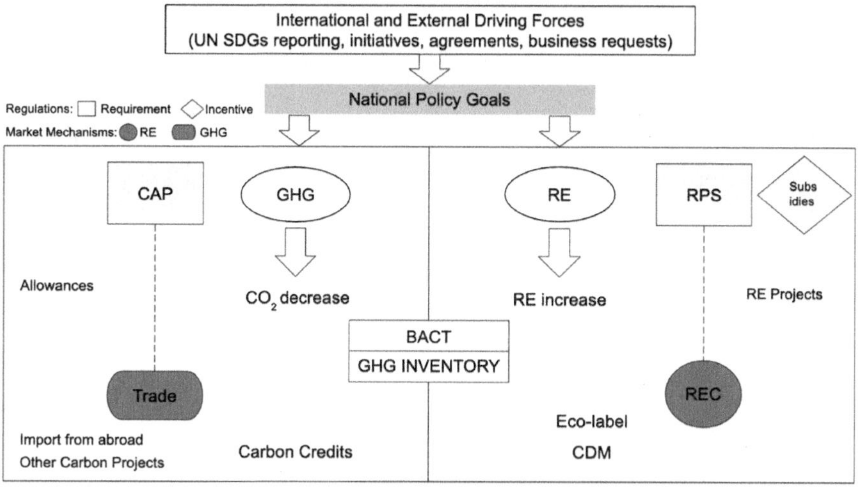

Figure 1.7. Consolidated climate markets mechanism analysis (CCLIMMA).

The issue of guaranteeing that investors have incentives to actually invest in mechanism transactions is not addressed by these laws, even though they give investors legal certainty that the mechanisms won't change. Investors will require assurance that Kyoto units will remain valuable after 2012 if considerable levels of mechanism activity are to be sustained. Therefore, rather than waiting until 2013, the issue of negotiating second commitment period targets, which support the financial worth of Kyoto units, must be addressed in the coming years. Similar language is used in the US Clean Air Act, which was passed to allow the federal government to still decide on permits when it was thought necessary to safeguard the public interest without having to worry about compensating polluters for potential violations of their legal rights. Despite the fact that the term "emissions" rather than the actual units created by the protocol is used in Decision 15/CP.7, the presence of this clause indicates that the Parties do not view ownership of Kyoto units as property rights. Instead, they view them as divisible representations of commitments made by sovereign states in the context of a multilateral agreement, which are thus subject to revocation, revision, and alteration through further negotiation.

This was done in response to concerns that the use of the Kyoto mechanisms might entrench as well as exacerbate existing emissions inequalities by encouraging Annex I Parties to seek reductions abroad. This clause will be considered for evaluating demonstrated progress in accordance with Article 3.2 of the Protocol. Every time the review procedure for national communications and supplemental information from AIPs under Article 8 of the Kyoto Protocol is finished. Putting domestic action first has moral and environmentally beneficial aspects. One of the most contentious areas of the post-Kyoto negotiations was the degree to which either fabricated cost-effectiveness concerns or legitimated quantitative limitations on the usage of the mechanisms. According to an agreement, a qualitative evaluation of whether Annex I Parties will satisfy the supplementarity condition necessitates that Annex I Parties submit information about their use of the mechanisms and domestic action as part of the information required by Article 7 that will be evaluated under Article 8. A second report on how each Annex I Party is demonstrating progress in accordance with Article 3.2 of the Protocol will also be included in the data pertinent to taking supplementarity into consideration. For the sake of clarity, it must be made clear that any implementation-related issues brought up by the qualitative assessment of supplementarity will be handled by the Facilitative Branch and not by the Enforcement Branch of the Compliance Committee. Although they are not supplementarity limits, it should be noted that the cumulative additions to a Party's allotted amount resulting from qualified land, land use change and forestry (LULUCF) projects under the CDM shall not, for the first commitment term, exceed 1% of that Party's base year emissions multiplied by 5(Kim, 2004).

1.9 FUNGIBILITY

The term "fungibility" refers to a variety of concerns including the type of first assigned amount each Annex I Party receives in accordance with the protocol, the interchangeability of Kyoto units, and the relationship between these two factors and the initial assigned amount. Parties are permitted to add and deduct ERUs produced through JI under Article 6 and AAUs produced through ET under Article 17 from their allotted amounts under Articles 3.10 and 3.11(Lawn, 2016)it is shown that the 6.2 % decrease in the Annex I group's emissions was largely made possible by the enormous decline in the greenhouse gases emitted by the Annex I (non-Annex II. The tCERs and lCERs produced by CDM LULUCF operations are equivalent to other Kyoto units in terms of compliance. Kyoto units can be exchanged for one another on a one-to-one basis regardless of how they were created since, according to Decision 19/CP.7, each of these units equals one metric tonne of carbon dioxide when computed using established global warming potentials (GWPs). These Kyoto units are comparable for compliance purposes. The distinct ability to bank Kyoto units for the following commitment period is another facet of fungibility. RMUs are not permitted to be banked at all, and Decision 19/CP.4 restricts the banking of CERs and ERUs to a maximum of 2.5% of a Party's allotted total. As a result, all Kyoto units other than AAUs are subject to banking limitations.

Regarding the legal status of assigned amounts and additions to them, a number of developing nations sought to clarify the legal status of the assigned amounts of Annex I Parties, and as a result, they pushed for the inclusion of language stating that the protocol does not create any rights, titles, or entitlements as stated above. Additionally, the Parties concurred that the designated amount of each Party should remain fixed for the commitment period once it has been entered into the compilation and database created by the Secretariat as part of the accounting mechanisms under Article 7.4. The allocated amount's fixed nature was also meant to alleviate the uncertainties of developing nations that the additions and subtractions of Kyoto units from the allotted amount would somehow weaken or change the nature of the legal obligations of Annex I Parties. According to Decision 15/CP.7, paragraph 6, these alterations will take effect without changing the quantifiable emission restriction and reduction commitments embodied in Annex B to the Kyoto Protocol(Gupta, Olsthoorn, & Rotenberg, 2003) developed countries are under a quantitative obligation to limit their emissions of greenhouse gases (GHG. The answer can be found in Decision 19/CP.7 on accounting modalities, which states that Kyoto units are not added to a Party's assigned amount until that Party designates those units to be used for purposes of fulfilling its commitment. This will be done at the conclusion of the commitment period. All Kyoto units were merely kept in a national registration. Therefore, CERs are as liquid as other Kyoto units prior to being designated for compliance use and can be freely transferred like the other Kyoto units. When this happens, CERs can no longer be deducted, but other units can be if the Party specified more units than were required to fulfill its pledge.

Decision 19/CP.7 strives to address the concerns of developing nations by defining the moment of addition in this manner while having no real influence on the transferability or liquidity of CERs. Stakeholder is defined in the JI and CDM modalities as the public, including persons, groups, or communities impacted, or likely to be affected by the project. The JI and CDM methods stipulate that specific types of information must be made readily available to the public in order to facilitate stakeholder participation in projects under these two initiatives. Such information is also required because there are several points during the JI and CDM project cycle where stakeholders may intervene to make sure that decisions, whether made by national authorities or international bodies, are in compliance with the modalities outlined in the various COP decisions. The success of the two project-based mechanisms, which if successful might result in thousands of initiatives around the world, depends on these points of intervention, the kinds of information that must be available, and the time of their availability. Although stakeholders aren't specifically mentioned in the regulations for emissions trading, they are in the regulations for the creation of national registries that will record transactions involving Kyoto units. These regulations stipulate that non-confidential information in national registries must be publicly accessible through an online user interface that enables interested persons to query and view the data kept in national registries. Therefore, NGOs and stakeholders, especially companies involved in the mechanisms, will play a crucial role in identifying the frequency, nature, and effects of any discrepancies that may appear, including whether the transaction log established by the UNFCCC Secretariat is actually operating as intended(Boiral & Heras-Saizarbitoria, 2017). Since Marrakesh, methods for building national registries have been progressively developed. These methods, as well as how issues involving sensitive data should be addressed, are still under consideration.

Another characteristic shared by all of the mechanisms is the emphasis placed in Articles 6, 12, and 17 on the fact that Annex I Parties may employ them to fulfill their Article 3.1 obligations. Even though participation is purely voluntary, once it is chosen, each Party must comply with some participation conditions. To guarantee that Parties maintain sovereign responsibility over transactions involving the mechanisms that take place within their jurisdiction, specific legal and administrative measures are outlined in each mechanism. All three processes give Parties the option to approve participation by private parties. In each situation, the modalities state that the Party that authorizes legal entities must continue to fulfill its Protocol duties and must make sure that their participation complies with those modalities. However, legal entities can only transfer and buy Kyoto units when a Party satisfies the rules for participation in the procedures. From an economic standpoint, it was thought important that entities participate in the mechanisms that fall under the purview of the Parties because, while mechanisms open only to governments are more effective than none at all, mechanisms limited to governments are less effective than mechanisms open to a wider range of actors. This is due to the fact that while government trading equalizes national marginal costs, it does not do the same for marginal abatement costs incurred by different sources inside each nation, and it is these latter costs that result in the greater reductions in compliance costs. All of the mechanisms specify the minimal environmental integrity requirements that Parties must meet in order to take part in the processes.

The permits or allowances produced by national emissions trading schemes established by nations that are not Parties to the Protocol, such as the US and Australia, will not be included in the Kyoto system. These non-Parties to the protocol are free to decide on their own whether or not to use Kyoto units for their domestic duties. The CDM modalities state that all Parties to the Convention may take part in CDM projects up until the protocol comes into effect. Non-Annex I Parties may only take part in the CDM if they are Parties to the protocol following the protocol's entrance into force. The CDM eligibility criteria for Annex I Parties pertain to eligibility at the time that such Parties will use CERs for compliance and are not as explicit. As a result, although the CDM modalities are primarily concerned with how Parties to the Protocol might engage in CDM activities, they neither rule out the possibility of non-Parties to the protocol taking part in CDM project activities nor provide guidance on how they may do so. All Parties to the Convention may participate in CDM project activities, the CDM Executive Board clarified at its ninth meeting, with reference to the validation requirements that must be examined by a designated operational institution prior to the entrance into force of the Kyoto Protocol. The registration of a planned CDM project activity can only thereafter take place, however, in accordance with the provisions of paragraphs 37(a) and 40(a) of the CDM modalities and procedures, once consent letters are acquired from Parties to the Convention that have ratified the Kyoto Protocol(Hickie, 2007).

Parties are required to nominate national authorities to oversee JI and CDM projects for both project-based procedures. In contrast to the CDM procedures, which call for all Parties to designate a national authority for the CDM, any Party interested in an Article 6 project shall advise the Secretariat of its chosen focal point. Developing nations pushed on the requirement to create a Designated National Authority (DNA) for the CDM because they wanted to make sure that governmental entities would be informed about CDM projects and efforts. One of the early lessons learned during the AIJ pilot phase was the direct project negotiations between Annex I countries and companies and non-Annex I parties, with governments frequently being kept out of the loop until a very late point. By February 2003, DNAs had been set up in over 50 developing nations and 10 Annex I Parties, with many more Parties close to finishing their internal procedures(Michaelowa & Jotzo, 2005). Amount allotted and pre-commitment period reports are established. The assigned amount under Articles 3.7 and 3.8 must be computed and recorded in accordance with the annex to Decision 19/CP.7 on modalities for the accounting of assigned amounts before any Annex I Party may engage in JI and ET transactions or employ CERs towards compliance. Each Annex I Party must submit a pre-commitment period report that includes all the data required to determine its assigned amount as well as other types of data required to show that the Party is capable of monitoring, tracking, and recording mechanism-related transactions, such as having a national system and national registry in place, in order to satisfy this requirement. This report must be turned in no later than 1 January 2007 or one year following the protocol for the Party's entry into force, whichever comes first. It is up for a careful examination by the review teams of experts.

1.10 GHG

A national system for the estimation of GHG emissions and removals in accordance with Article 5.1 of the Kyoto Protocol must be in place by no later than 1 January 2007 for an Annex I Party to engage in JI and ET transactions or utilize CERs. The institutional, legal, and administrative setups made by an Annex I Party to be able to properly estimate, report, and store GHG inventory data are referred to as national systems. The Marrakesh Accords established guidelines for the development of national systems and review procedures with the goal of ensuring that they are reliable enough to satisfy mechanism criteria. These guidelines mandate that Annex I Parties abide by the 1996 IPCC Revised Guidelines for National Greenhouse Gas Inventories, which incorporate standard operating procedures and reporting formats developed by the IPCC and later approved by the COP for use by Parties when calculating emissions data and assembling their annual GHG inventories(Lim et al., 1999). In accordance with the requirements for national systems, Annex I Parties must appoint a national authority that is responsible for maintaining a national system, including the implementation of quality assurance/control processes to guarantee that GHG inventories meet the standards established by Kyoto. The pre-commitment period report, which serves to determine whether Parties have the administrative ability to track mechanism transactions and to monitor their GHG emissions, must include a description of the national system. The national system component of the mechanism participation criteria is anticipated to cause problems for many EITs since they lack the capacity to submit annual inventories in accordance with IPCC guidelines. This is unless further capacity-building initiatives are done.

The phrase national registry refers to the protocols and systems established by an Annex I Party to guarantee the precise tracking of transactions under the Kyoto process and LULUCF accounting under Articles 3.3 and 3.4. The UNFCCC Secretariat will keep an independent transaction log in addition to a separate CDM registry as part of a larger accounting framework that includes national registries. Registries and their related elements are essential for trade under the Kyoto mechanisms as well as connecting the Kyoto mechanisms with independent state programmes. The UNFCCC Secretariat must concentrate on creating the independent transaction log, and Annex II Parties must support this work, according to SBSTA-19's conclusions, which were approved alongside the COP's technical instructions to help Parties create registries that are compatible. Whether or not they decide to take part in the mechanisms, all Annex I Parties must designate an organization to act as the administrator of the national registry that must be created and kept by all Annex I Parties as part of the accounting modalities required to track their assigned amounts. This is a requirement of the protocol. Since registries play such a significant role in the administrative framework for monitoring transactions under the mechanisms, expert review teams will carefully examine and test their effectiveness as part of the pre-commitment period report and review procedures. The evaluation of the pre-commitment period report, which is required for Annex I Parties to establish their allotted amounts, will include verification that they are functioning successfully.

For many Annex I Parties, especially several EITs that continue to struggle with capacity-related issues in completing comprehensive inventories, the application of modifications will be a crucial component. Applying modifications to complete missing or insufficiently justified emissions data is expected to occur frequently, if present inventory standards give any indication. The Enforcement Branch of the Compliance Committee will decide on the case when an ERT proposes an adjustment and an Annex I Party disputes it. In an effort to reduce the number of cases brought before the Enforcement Committee involving disputed adjustments, the COP-9 adopted technical guidance in the form of recommendations for COP/MOP on methodologies for the application of adjustments by ERTs. These recommendations ensure that the conditions in which these are proposed and applied are understood and acceptable to Annex I Parties. In accordance with

Decision 20/CP.9, the Secretariat was tasked with developing a procedure that would allow ERTs to gain experience with changes throughout the inventory review process in 2003–2005 using actual inventory data, with the approval of the Party in question. Taking into consideration the methodological work by the IPCC to provide Good Practice Guidance for LULUCF, the topic of guidance on how adjustments will be implemented in respect of sink sources were reviewed at COP-10(Höhne, Wartmann, Herold, & Freibauer, 2007). It is significant to note that for the first commitment period, Parties will be required to submit annual sink inventory data as part of their mechanisms eligibility requirements. This is made plain in the mechanisms' eligibility provisions. However, since the quality evaluation for establishing eligibility is only applicable to the portions of the inventory pertaining to emissions of GHGs from sources/sectors specified in Annex A to the protocol, the quality of sink data will not matter. Because many Annex I Parties, especially EITs, still struggle to provide correct data on sinks, the lack of a quality evaluation of sink data was a key problem at Marrakesh. The fact that Parties who fail to provide accurate data on sinks will only be allowed to issue RMUs in respect of sinks on which they have reported adequately, a provision that provides some environmental integrity but does not create incentives for the general improvement of sinks data as a whole had to satisfy Parties that had wanted to link improvement of sinks data to mechanism eligibility.

All three procedures involve the submission of supplemental information on assigned amounts by Annex I Parties in accordance with Article 7.1 and compliance with rules for the accounting of assigned amounts adopted in accordance with Article 7.4. The KP reporting criteria outlined in Decisions 22/CP.7 and 22/CP.8, as well as Decision 19/CP.7, specify the additional information that must be included in national communications(Massai, 2011). Annex I Parties efforts to carry out its obligations under Article 3.14 relate to questions of implementation, and are not a requirement for qualifying for the mechanisms. In Decision 18/CP.7 on emissions trading, it is stated that a commitment period reserve (CPR) must be created and kept in place. A Party must abide by the provisions pertaining to it in order to engage in mechanism transactions in accordance with Decision 18/CP.7, even though it is not expressly stated as a prerequisite for eligibility for all mechanisms. In accordance with Article 17, the CPR is explored more below under IET. However, a number of JUSCANZ nations were unwilling to agree to accept the compliance procedures as an eligibility requirement for using the mechanisms due to unresolved differences regarding the binding nature and form of the compliance procedures that should be implemented by the COP/MOP. It is crucial to stress that the Enforcement Branch's inspection of the agreed eligibility standards, which all Parties believe is required, was not meant to be undermined or affected by this phrasing. In accordance with Decision 24/CP.7, the Enforcement Branch of the Compliance Committee will be in charge of monitoring the eligibility requirements for the mechanisms.

The fact that the IET modalities reached in Marrakesh, after four years of negotiations, are barely two pages long, serves as sufficient proof that, if basic principles can be agreed upon, the conceptual simplicity of emissions trading is attractive. However, it also requires binding targets, thorough reporting, and a strong national and international infrastructure to monitor, track, verify, and enforce compliance in order to execute. Article 17 makes it possible for an Annex I Party to find itself in violation of its Article 3 commitments through deliberate or unintentional overselling, especially where there is lax international enforcement of compliance with international commitments. This is because selling Kyoto units through emissions trading is likely to prove profitable(Manley & Maclaren, 2012). The commitment period reserve (CPR) idea was first put up by OASIS as a compromise to deal with large-scale selling by defining the minimum number of Kyoto units a country must ever hold in its national registry and therefore reducing the potential for non-compliance. Setting the CRP standards in a way that would protect against non-compliance but not being excessively difficult as to reduce market liquidity was one of the issues. Any Kyoto units that are still valid for that commitment period may be included in the reserve. The adopted restrictions would stop widespread non-compliance.

At Marrakesh, several Parties insisted that the word "should" did not make maintenance of the CPR at 90% levels mandatory, which led to the CRP-related text in Decision 18/CP.7 being moved from the eligibility section into a later section of the text. Negotiators decided to maintain the Bonn wording but added that a Party shall not make a transfer which would result in these holdings being below the required level of the commitment period reserve(Schulze, Valentini, & Sanz, 2002)Bonn and Marrakesh came to a final Agreement on the Kyoto Protocol, which is thus ready for ratification by the individual nations. The Agreement was only achieved by allowing countries to offset their fossil fuel emission targets (on average 95% of the 1990 emissions. This was done because the absence of a mandatory level would have offered no protection against overselling and undermined the spirit of the compromise reached in Bonn. An Annex I Party is required by Decision 19/CP.7 to determine its allotted amount after determining the level of its CPR. The transaction procedures outlined in this Decision also make it challenging for any Annex I Party to violate the CPR's minimum level without drawing immediate attention because the transaction log that the Secretariat will keep track of will confirm that all transactions involving Kyoto units comply with the necessary regulations. The log will alert the relevant national registry, which is required by law to halt the transaction, if any transaction is shown to be out of order, for instance, because it exceeds the CPR restrictions. Additionally, the acquiring registry will be informed. Any units that have already been transferred in violation of the CPR restrictions will be considered invalid for compliance reasons up until the level of the CPR is reestablished, with the Parties having 30 days to bring the CPR back into compliance. As part of the review procedure for the Party or Parties involved as per Article 8, the Secretariat must be informed of all discrepancies. Any Party that permits its CPR to drop below necessary levels without taking swift corrective action runs the possibility of being brought before the Enforcement Branch and being prevented from transferring Kyoto units.

An Annex I Party is permitted to place restrictions on who it will trade with under Article 17 and with regard to what because participation in Article 17 is voluntary(Sharkey et al, 2020)several families of brominated flame retardants (BFRs. For instance, a Party is allowed to choose the kind of ERUs and CERs it will accept as payment for its responsibilities and the number of CERs its legal companies can purchase in order to fulfill any domestic requirements it may impose. Because transfers of Kyoto units represent exchanges of sovereign commitments that are not covered by international trade law disciplines because these deal with trade restraints relating to conventional types of goods and services, carbon permits cannot be regarded as goods and services, there would appear to be no trade law implications from these types of restraints. The establishment of a national emissions trading plan is not required for parties who are eligible to take part in Article 17 emissions trading. Even in the absence of a national trading scheme, they may authorize governmental agencies or other legal entities to participate. National trading schemes will, nevertheless, be established by many Parties as a type of internal policy action.

Lack of quantifiable mitigation pledges by non-Annex I Parties in the CDM context encourages individuals involved in CDM projects to artificially inflate the number of CERs claimed, for example by manipulating counterfactual baseline scenarios. Therefore, multilateral control is needed for the procedures leading to CER issuance. The Executive Board of the

CDM (EB), which is responsible for providing this oversight, consults with independent organizations to evaluate how well CDM project activities adhere to internationally recognized norms. An additional ground-breaking feature of the CDM is that a portion of the funds from certified project activities are automatically deposited in the CDM Registry, which is run by the UNFCCC Secretariat, to support climate change adaptation in developing nations at risk as well as, eventually, to pay for CDM-related administrative costs. In addition to producing CERs, one strategic justification for the CDM's inclusion in the protocol was to give non-Annex I Parties a quantified way to contribute to mitigation commitments and to learn more about trading mechanisms without requiring such Parties to adopt legally binding mitigation targets. Thus, participating in the CDM is a part of the larger initiatives of poor nations to mitigate global warming in a way that respects their aspirations for sustainable development and allows for "learning by doing." There were insufficient operational specifics in the CDM project cycle Article 12 of the Protocol defining the roles of different participants in CDM project activities. There were numerous project cycles with various implications, as well as numerous conflicting visions and preferences.

CHAPTER 2

Carbon Footprint

CONTENTS

2.1 INTRODUCTION

Sustainability (Figure 2.1) is about stabilizing the currently destabilizing interaction between Earth's two most complex systems i.e., human culture and the living world, writes Paul Hawken, a widely respected author on business and the environment(Clark & York, 2005).

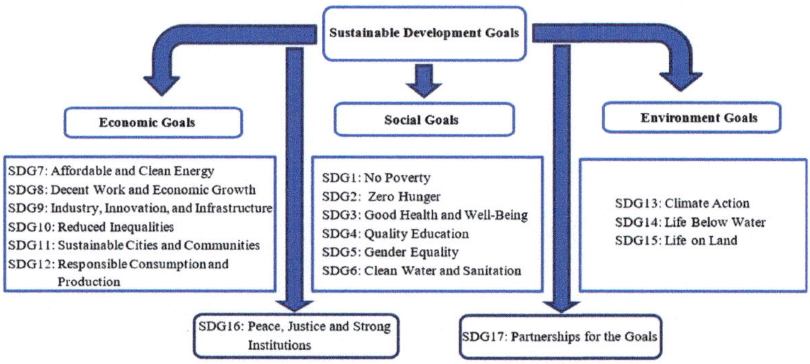

Figure 2.1. Sustainability. (Source: Triki R, Maâloul MH, Bahou Y, Kadria M. The Impact of Digitization to Ensure Competitiveness of the Ha'il Region to Achieve Sustainable Development Goals. *Sustainability*. 2023; 15(2):1661. https://doi.org/10.3390/su15021661, Creative Commons License)

In order for human growth and the ecological system of the Earth to coexist and work in harmony without degrading or destroying one another, sustainability aims to bring them into alignment. The impact of humans on the climate is arguably the biggest and most obvious disconnect between human growth and the Earth's ecological system today. While there are many other environmental issues, climate change and its effects on the Earth's temperature are among the most urgent sustainability concerns of our generation. Since the early 1990s, the dangers of climate change have drawn more attention from credible scientists and well-known supporters, such as former U.S. vice president Al Gore. Many scientists believe that as Earth's average temperature increases, sea levels will rise, violent hurricanes will become more frequent and intense, there will be widespread flooding and droughts for extended periods of time, there will be a decrease in the amount of food available globally, and there will be a higher rate of species extinction. Scientists have linked rising energy use and greenhouse gas emissions, particularly carbon dioxide (CO2), from human activities to the phenomenon of climate change. The burning of fossil fuels and different industrial/manufacturing processes that release GHGs are at the core of these human activities.

Figure 2.2. Example of an instructional website-Climate Communication.

The facts and problems around climate change are still largely unknown to the public. Many activist and instructional websites do a great job of simplifying the facts and involving the audience; both beginning and experienced readers should check them out. There are different sections on what is happening to our climate, how it will affect us, and what we can do that are data- and science-based. For example, http://climate.nasa.gov; http://epa.gov/ climatechange/, the National Aeronautics and Space Administration (NASA) and the United States Environmental Protection Agency (USEPA) both have science-based educational websites on climate change with links specifically for the general public, children, or educators. The Post Carbon Institute, a nonprofit organization (http://www.postcarbon.org/), strives to support the transition to a world where our civilization is independent of hydrocarbon fuels and does not release large amounts of carbon dioxide into the atmosphere. Two websites that offer information have been particularly successful at informing and attracting public attention. The CO_2Now website (http://co2now.org/), for instance, makes it simple to view the most recent CO_2 level and its effects(DeStefano, 2015). The atmospheric CO_2 concentration was 389.00 parts per million (ppm) in September 2011, up from 384.79 ppm and 386.80 ppm in September 2009 and September 2010, respectively. The 350 website (http://www.350.org), another website, attempts to create a worldwide grassroots movement to address the climate issue. It briefly discusses how preindustrial CO_2 concentration was 275 ppm, how it has been increasing at a rate of roughly 2 ppm annually, and how the most recent scientific research indicates that we must reduce it to at least 350 ppm in order to maintain a safe and largely stable planet Earth. Unfortunately, since 1988, the atmospheric CO_2 level has been above 350 ppm, and no political action has been taken to lower it.

2.2 CARBON DIOXIDE

Our civilization depends on the atmospheric CO_2 concentration since it affects how much the global systems will be negatively impacted. The Earth's temperature increased by 0.74°C over pre-industrial levels in 2005, and the atmospheric CO_2 concentration was 379 ppm (Intergovernmental Panel on Climate Changc [IPCC] 2007)(Lenton & Vaughan, 2009). For CO_2 concentrations up to 790 ppm, the sea level is projected to rise by 0.4 to 3.7 m, and the temperature is projected to rise by 6.1 °C above pre-industrial levels. The 70 percent of the world's population that lives in coastal regions would be devastated by this sea level increase(Neumann, Vafeidis, Zimmermann, & Nicholls, 2015). Model predictions indicate severe extinctions worldwide if global average temperatures rise by more than 3.5°C from pre-industrial levels (IPCC 2007). It was found that 2°C was the maximum safe temperature to prevent major economic and ecological consequences (IPCC 2007). Because the challenges of the environment and energy are interrelated, climate change is frequently mentioned in relation to energy use. On the one hand, burning fossil fuels increase the atmospheric GHG concentrations. It is anticipated that the availability of fossil fuels would drastically decrease within the lifetime of the current generation, leading to severe energy problems. Therefore, cutting back on fossil fuel use will help us manage and adapt to both the problems of climate change and the energy crisis. Many organizations are now making it a point to adapt to a world without fossil fuels. This incentive stems from the finite supply of fossil fuels, which are predicted to peak or have peaked in the twenty-first century, as well as the environmental issues associated with burning them.

The Hubbert peak theory(Figure 2.3),put forth by American geophysicist King Hubbert, provides the scientific justification for this peaking and decline of fossil fuel supply in current and future generations(Priest, 2014). Since all fossil fuels are limited in supply and non-renewable within the time frame of human civilization, the Hubbert peak theory has been most frequently used, but it also applies to peak natural gas and coal. The term "post carbon" is occasionally used to describe a world free of fossil fuels and to support the present transition to such a world. The term "post carbon" has gained popularity. Our society is already taking action to combat climate change and make preparations for a post carbon future. These projects are frequently linked to bigger sustainability ones. For instance, local food movements and local campaigns work to support local economies while simultaneously advancing practices that are frequently less dependent on fossil fuels and intensive energy consumption.

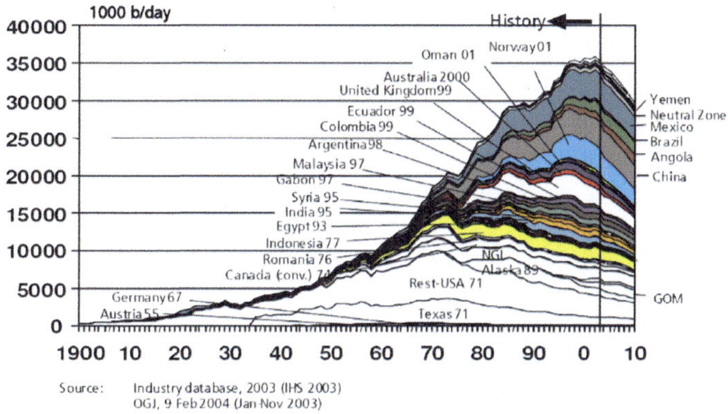

Figure 2.3. Hubbert peak theory. (Source: https://upload.wikimedia.org/wikipedia/commons/thumb/1/1d/Hubbert_world_2004.svg/1200px-Hubbert_world_2004.svg.png, Creative Commons License)

The demand for fossil fuels decreases when food and goods are transported across shorter distances. Similar to this, local food systems frequently rely less on pesticides made from fossil fuels. These incidents highlight the significance and requirement for accurately measuring and reducing energy and GHG emissions. While there are many materials available on climate science and addressing climate change is going to be one of the most critical challenges of our generation, a systems approach on how to assess and manage GHG emissions, particularly for businesses, has been lacking in the literature.

Many analytical models have been created to assist businesses in reducing their use of energy and greenhouse gas emissions. More than 20 businesses have used this strategy, and it has produced results in terms of the environment and the economy. The model is predicated on the saying "If you can't measure it, you can't manage it" in business. The model aims to achieve the following objectives by quantifying energy and GHG emissions: Prioritize areas for emission reductions and efficiency improvements by identifying significant sources of emissions, Start preparing for a world without carbon by lowering GHG emissions and the organization's dependency on fossil fuels, Improved raw material utilization and lower energy costs will boost business competitiveness, Enhance corporate image as businesses become green, Reduce pollution by consuming less energy and utilizing clean, renewable energy sources. The main objective of this kind of work is to enable businesses and organizations to lower their operational expenses and environmental impact by limiting the energy they consume and the GHGs they release. This entails cutting down on GHG emissions, energy use, transportation, and storage, reducing environmental taxes, and controlling pollutants to enhance environmental quality(Ekins & Speck, 1999). Many businesses can boost their profits by cutting back on their energy bills, raw material purchases, and other running expenses. However, due to time restrictions and a lack of expertise in the area, many businesses lack the ability to complete an audit of their energy and carbon footprint (Figure 2.4).

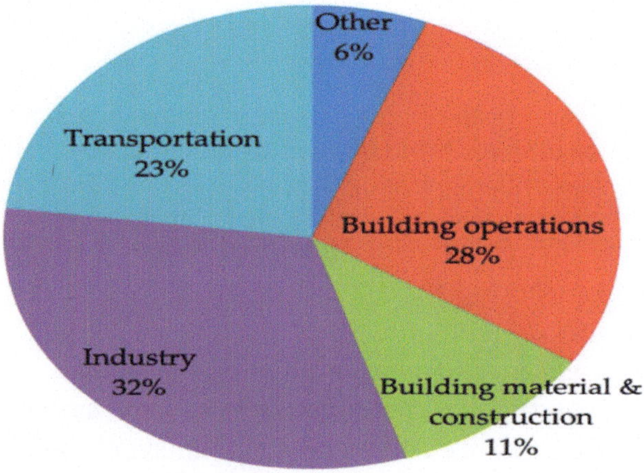

Figure 2.4. Carbon footprint. (Source: Ahmed Ali K, Ahmad MI, Yusup Y. Issues, Impacts, and Mitigations of Carbon Dioxide Emissions in the Building Sector. *Sustainability*. 2020; 12(18):7427. https://doi.org/10.3390/su12187427, Creative Commons License)

People frequently have different expectations when they hear several of the common terminologies used in the environmental industry. Finding common definitions for these phrases can be difficult because numerous businesses and government organizations develop their own abbreviations, frequently by combining technical and operational elements. It is crucial to define carbon footprint analysis first. The measuring of GHG-emitting processes, their sources, as well as their composition and quantities, is known as a carbon footprint analysis. To establish the net emission rates, a carbon footprint study should also take into account the GHG sinks and removal rates. The term carbon footprint is generally used in a vague way to describe how much GHG is emitted as a result of an individual's or an organization's activity. Estimating the carbon footprints of countries, goods, or services is also one option. The terms carbon footprint analysis and greenhouse gas inventory are interchangeable. The term carbon is chosen since CO_2 is the main GHG released by human activity. Methane (CH4) and nitrous oxide (N_2O), two other GHG emissions, also significantly contribute to global warming(Wulf, Maeting, & Clemens, 2002)nitrous oxide (N2O. Emissions from these other gases are normalized to the mass of CO_2 in order to have a single unit for reporting results, and the carbon footprint results are presented as the mass of CO_2 equivalent (CO_2e). Since the majority of emissions from a person, a product, or an organization are caused by the use of fossil fuels for energy, the terms energy footprint and carbon footprint are frequently used interchangeably. Technically speaking, though, this is untrue because GHGs can come from other sources besides fossil fuels. Other sources, including industrial operations, agriculture, land use, land-use change, forestry, and waste, may be useful for cataloging greenhouse gases and computing the carbon footprint, depending on the characteristics of the business or product.

2.3 CARBON FOOTPRINT

Analyzing a company's carbon footprint frequently involves an audit or assessment process. This calls for the tracking and measurement of data as well. Reducing energy use, changing processes or equipment, and reallocating resources are all ways to reduce GHG emissions and the energy footprint. The energy and carbon footprint audit normally entails, from a corporate perspective: Researching the company's processes and total energy and GHG generation; Developing recommendations to enhance process efficiency and reduce energy and GHG emissions and energy expenses. Greenhouse Gases GHGs are atmospheric gases that absorb and emit thermal infrared radiation.

These GHGs (Figure 2.6) are all naturally present on Earth. Six types of gases that are part of the Kyoto Protocol are commonly taken into consideration in carbon footprint analyses: CO_2, CH_4, N_2O, and other anthropogenic gases like sulfur hexafluoride, hydrofluorocarbons (HFCs), and perfluorocarbons (PFCs)(Al-Ghussain, 2019).

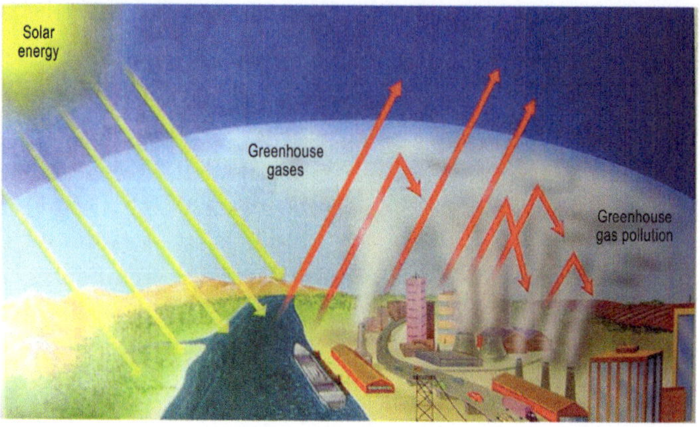

Figure 2.5. Greenhouse effect. (Source: Thacker I, Sinatra GM. Visualizing the Greenhouse Effect: Restructuring Mental Models of Climate Change Through a Guided Online Simulation. *Education*

It is noteworthy to note that although the idea of global warming has been around for more than a century, it has only recently gained popularity. The idea that human activity and the subsequent combustion of fossil fuels may cause an increase in global warming was first proposed in 1896 by a Swedish scientist by the name of Svante Arrhenius(Weart, 1997). He insisted that there was a connection between temperature and atmospheric CO_2concentrations. The topic received little attention from the general public until the 1980s, when scientists noticed an increase in the yearly mean temperature of the Earth.

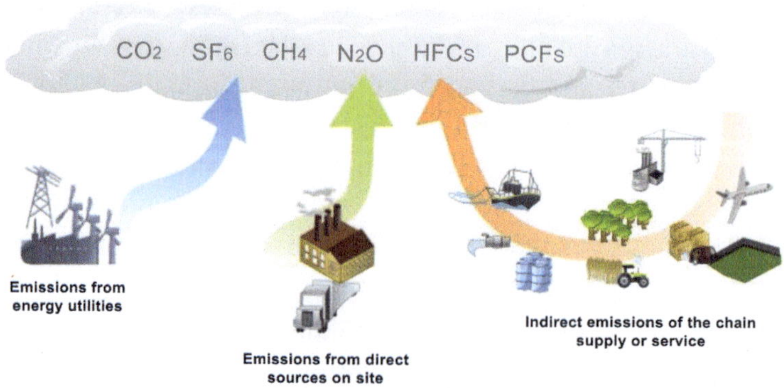

Figure 2.6. Sources of main greenhouse gas emission.

The IPCC has more than 2000 experts from a wide range of disciplines on staff. The IPCC, which is regarded as the top organization in the field of

GHG emissions and global warming, regularly publishes climate change reports. Since the middle of the 20th century, Earth's near-surface air and oceans have warmed on average; this trend is expected to continue(Stott et al., 2000). In 2006, the average surface temperature of the Earth was 14.5 °C (IPCC 2007). This temperature is an estimation based on both land and water surfaces. According to the National Oceanic and Atmospheric Administration-National Climatic Data Center (NOAA-NCDC) (2011), the global mean annual surface temperature estimates for the last century (from 1901 to 2000) were 13.9°C for land and sea surface combined, 16.1°C for sea surface, and 8.5°C for land surface(Trenberth & Fasullo, 2013). Since the beginning of the industrial age, the earth's surface has generally warmed, however, land areas have warmed more quickly than oceans (IPCC 2007).

The United Nations Intergovernmental Panel on Climate Change (IPCC) (Figure 2.7) is the top scientific authority on climate change in the world. The latest global scientific, technical, and socioeconomic data essential to understanding climate change are examined and evaluated by IPCC. The most often used and quoted sources on climate change are the IPCC assessment reports. In 1990, 1995, 2001, and 2007, consecutively, the IPCC released its First Assessment Report (FAR), Second Assessment Report (SAR), Third Assessment Report (TAR), and Fourth Assessment Report (AR4).

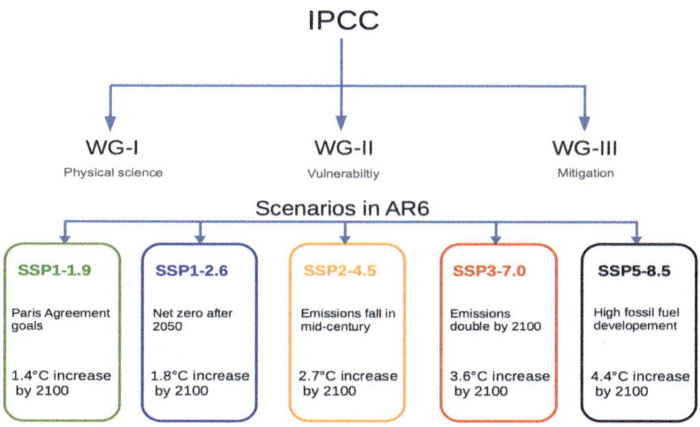

Figure 2.7. United Nations Intergovernmental Panel on Climate Change (IPCC) .

According to AR4, the linear warming trend over the past 50 years has been 0.13°C (0.10 to 0.16°C) each decade (IPCC 2007). In the instrumental record of the global surface temperature (since 1850), 11 of the last 12 years (1995–2006) were among the 12 warmest years (IPCC 2007). Scientists at

Goddard Institute for Space Studies (GISS)of NASA reported that 2010 and 2005 were the warmest years on record for global surface temperatures (NASA GISS 2011). Since the middle of the 20th century, the majority of the observed temperature increase has been attributed to rising GHG concentrations, which are brought on by human activities such as fossil fuel combustion and deforestation(Wuebbles & Jain, 2001).

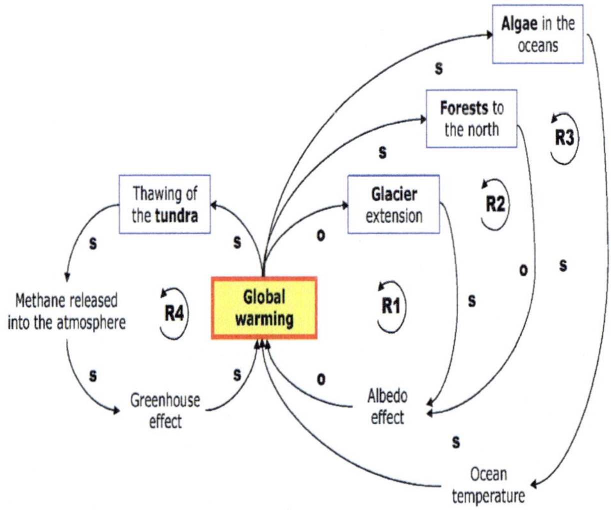

Figure 2.8. Global warming.

2.4 GLOBAL WARMING

Although there is cause for concern regarding global warming, (Figure 2.8) the rise in temperatures on their own is not the main cause. The global climate changes as a result of global warming, which has an unfavorable impact on the overall physical, chemical (pH), and biological (species adaptation) systems in ways that are irreversible and outside of the ranges on which our current civilization depends.

Therefore, what is of concern is how global warming is affecting Earth's systems and how these systems are affecting human civilization. Because of this, terminology like climate change and global change are better suited to describing the issue and offering solutions than the term global warming. GHGs contribute to global warming, which in turn causes climate change and, ultimately, global change. The increase in Earth's surface temperatures is referred to as global warming. The term climate change describes how these elevated temperatures have altered the global climate. Among these changes are variations in precipitation patterns, the availability of freshwater, the extent of glaciers and sea ice, ecosystem health, and biodiversity. The word global change is the most inclusive and applies to all types of global change, some of which are caused by factors like population, globalization, economy, and pollution and others which are caused by factors like sea levels, ocean pH, and biodiversity.

2.5 LCA

A life cycle assessment (LCA), often referred to as a life cycle analysis is an examination and assessment of the environmental effects of a specific good or service brought on by or required by its use(Rebitzer et al., 2004) as well as other interventions (e.g., land use. The goal of LCA is to identify the least damaging option by evaluating a good or service based on potential environmental problems associated with its production, use, and disposal.

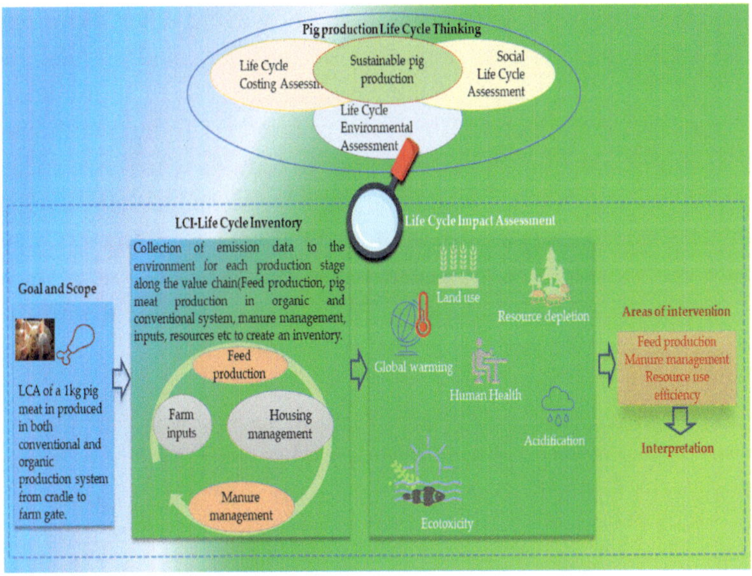

Figure 2.9. Life Cycle Assessment.

Life Cycle Assessment (LCA)(Figure 2.9) is a technique for taking into account the impact and results of technology chains that are involved in the production and consumption of products and services. The phrase life cycle alludes to the idea that a fair, comprehensive evaluation necessitates an analysis of raw material production, manufacture, distribution, use, and disposal, as well as all interim transportation steps required by or contributing to the existence of the product. The life cycle of the product is made up of all those periods or steps. The evaluation of this life cycle offers a thorough method for comprehending the effect of the product. The LCA idea can be applied to either improve a company's environmental performance or the environmental performance of a particular product (eco-design). LCA allows for the analysis of numerous environmental impacts. Human health and natural resources are examples of end-point consequences, while climate change, acidification, and eutrophication are examples of intermediate impacts. The life cycle cost analysis (LCCA) and social life cycle analysis (SLCA) frameworks can now be used to incorporate social and economic implications into the LCA framework.

With its consideration of the three pillars of sustainability (economic, social, and environmental), life cycle sustainability analysis (LCSA) has most recently become the main sustainability tool. Life cycle sustainability impacts come in a variety of forms. The analysis of the carbon footprint simply quantifies the effects of climate change; however, other environmental, social, and economic effects may also be pertinent and significant when making decisions about a particular issue or analysis. The methodologies used to quantify the impacts in life cycle research are fraught with numerous unknowns. However, the global warming potential, which is a measure of GHG emissions, may be the impact that can be assessed with the least amount of uncertainty due to the presence of more databases. Emissions from various GHGs are normalized to CO_2 equivalence and expressed as mass of CO_2 equivalence in the category of the potential impact of global warming. The possibility of global warming is probably the most often studied category of life cycle impact in recent years. An important methodological framework for carbon footprint analyses is provided by LCA's usage of the category for the possible impact of global warming. However, we can reiterate that a more thorough LCA will produce more results than only a carbon footprint analysis. The LCA is made to examine all pertinent effect categories. In other words, the environmental effects of a particular good, service, or activity generally include more than just global warming. They may also include effects on eutrophication, acidification, human toxicity, terrestrial toxicity, or land use, among other things All pertinent environmental impacts should be measured in order to make a more educated decision. Focusing solely on carbon footprint analysis at the expense of ignoring other effects may offer a constrained viewpoint and, as previously mentioned, lead to the burden being shifted, for instance, from global warming to eutrophication.

Although the LCA methodology was created for the analysis of goods or services, it is theoretically possible to estimate the carbon footprint of any entity, including an individual, an industry, a company, a community, or a whole country. When LCA and the category for global warming impact are utilized, and the study leaves out the operation and end-of-life phases, the phrase "carbon footprint" may occasionally be replaced with "embedded carbon." Embedded carbon is a measurement of the total amount of greenhouse gas emissions that have occurred at each stage of the production process, starting with raw material extraction and ending with the final product(Weir & Muneer, 1998). The total amount of energy utilized to create a product is also referred to as embedded energy. Although the LCA approach can potentially be used to analyze anything (organizations, industries, or communities), it has mostly been designed for and used for this purpose. This is due to two factors. First, LCA monitors emissions from every stage of the target entity's life cycle. The life cycle phases of items may be easily identified, and once these phases are known, the emissions from those products can be calculated. However, the life cycle phases for individuals, groups, businesses, or communities are too lengthy and variable to be identified, making them irrelevant or impossible. Second, LCA monitors all emissions, including direct ones, those resulting from power and other upstream and downstream processes as well as other indirect emissions. The precise quantification of indirect emissions from upstream and downstream processes is exceedingly challenging(Lee, 2011).

It becomes increasingly difficult to measure these indirect emissions when the entity under study becomes more complex, as there would be far too many emissions to keep track of. For goods or services whose borders and interactions with other entities are well understood, these emissions can be quantified and modeled more simply. A more focused strategy that evaluates only the emissions for a specific time period, usually one year, can be used in situations where the LCA framework is no longer relevant or cannot be simply applied to estimate the carbon footprint of a person, business, or organization. As a result, a company's carbon footprint refers to the annual emissions generated by its operations. The main methodology employed in such measurements is the GHG Protocol. These studies track both the direct and indirect emissions from electricity use. It is optional to measure indirect upstream and downstream emissions. LCA is a young but rapidly expanding scientific field. The earliest studies on product comparisons were conducted in the 1960s and 1970s, which is when LCA principles initially emerged. The International Organization for Standardization (ISO) and SETAC standardized the methodologies of the LCA in the 1990s when it first gained international acclaim. LCA techniques underwent additional development throughout the 2000s. LCA studies also grew in number steadily. Three more globally significant resources, water, nitrogen, and ecosystems have all been used, and the footprint concept has been applied to them.

The ideas behind the nitrogen and water footprints are comparable to those behind the energy footprint. They serve as indicators for the direct and indirect uses of an environmental indicator, in this case, reactive nitrogen and freshwater. Because freshwater water quantity and quality issues are frequently covered in the media and because most people are aware of their own basic water demands, the water footprint idea and its significance are reasonably simple to comprehend. The public is mainly unaware of the relevance and significance of the nitrogen footprint concept, in contrast to the quantity of popular news on carbon and water footprints, which has received far more attention. The global nitrogen cycle disturbance caused by humans is best understood using the quantitative indicator known as the nitrogen footprint. The substantial conversion of unreactive nitrogen (N_2) by humans to more reactive nitrogen (e.g., N_2O, nitrate [NO^{3-}], nitrite [NO^{2-}], ammonia [NH_3], and ammonium [NH^{4+}]) is to blame for the disturbance of the global nitrogen cycle. The environment and human health have suffered as a result of this changeover. Prior to the industrial revolution, the amount of reactive nitrogen produced by humans and released into the environment was negligible(Erisman et al, 2011). The production of reactive nitrogen by anthropogenic processes increased tenfold between 1860 and the early 1990s as a result of the burning of fossil fuels, biological nitrogen fixation in human-cultivated land, and the Haber-Bosch process' conversion of atmospheric, nonreactive N_2 to ammonia fertilizer. The Haber-Bosch process increased food production, enabling the existence of half of mankind; nevertheless, it also had significant effects on the nitrogen cycle by raising N_2O and NO^{3-} concentrations in soil and water. Major environmental issues result from these elevated concentrations, including a worsening of the greenhouse effect, a reduction in the ozone layer, a rise in smog, an increase in acid rain, and water contamination. In order to properly manage the global nitrogen cycle (Figure 2.10) and take on this major challenge, the nitrogen footprint concept and related technology are crucial.

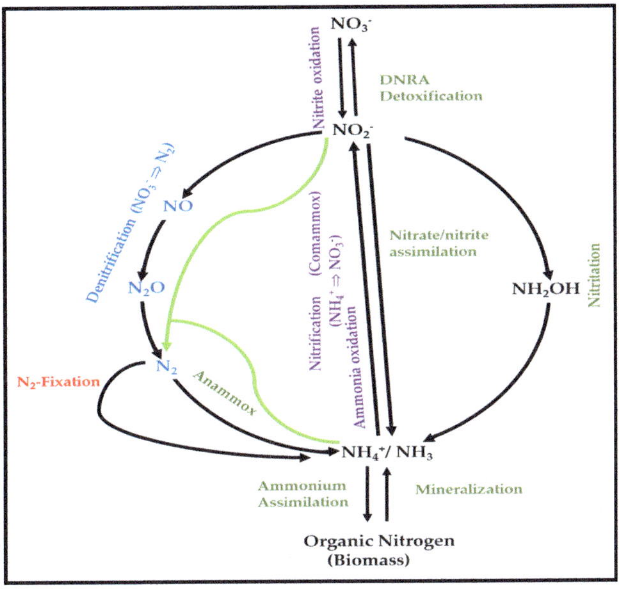

Figure 2.10. Global nitrogen cycle. (Source: Martínez-Espinosa RM. Microorganisms and Their Metabolic Capabilities in the Context of the Biogeochemical Nitrogen Cycle at Extreme Environments. *International Journal of Molecular Sciences*. 2020; 21(12):4228. https://doi.org/10.3390/ijms21124228, Creative Commons License)

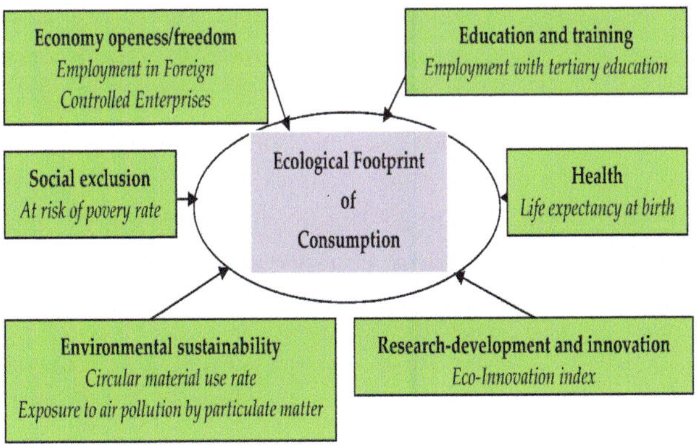

Figure 2.11. Ecological footprint. (Source: Ghita SI, Saseanu AS, Gogonea R-M, Huidumac-Petrescu C-E. Perspectives of Ecological Footprint in European Context under the Impact of Information Society and Sustainable Development. *Sustainability*. 2018; 10(9):3224. https://doi.org/10.3390/su10093224, Creative Commons License)

2.6 ECOLOGICAL FOOTPRINT

The ecological footprint (Figure 2.11) is interpreted somewhat differently from the water, carbon, energy, or nitrogen footprints(Matuštík & Kočí, 2021) (Matuštík & Kočí, 2021). An individual, population, or activity's ecological footprint measures how much biologically productive land and water is needed to produce all the resources it consumes and to absorb the carbon dioxide emissions it produces. The ecological footprint is a measure of human demand from nature. According to data from 2007, the global ecological footprint was 1.5 times greater than its biocapacity, indicating that 1.5 planet Earths would be required to meet human needs from nature as they exist today. We currently have a 0.9 global hectares per capita ecological deficit because more is being sought than the Earth can sustain. Overshoot is the term for this imbalance, which shows that we are using up more of the planet's natural resources than it can sustain on a yearly basis. The U.S. has bigger per capita footprints than the rest of the world in terms of carbon, water, and ecological footprints. They established limits for nine Earth-system processes, including chemical pollution, ocean acidification, global freshwater consumption, stratospheric ozone depletion, rate of biodiversity loss, climate change, and rate of aerosol loading. The borders of three of these nine Earth systems, climate change, biodiversity loss, and nitrogen cycle have already been crossed, and footprint calculators are now available for each of these issues. The Biodiversity Indicator has chosen the ecological footprint as a biodiversity indicator even though it does not directly measure biodiversity(Feest, 2013). This is because it measures the level of ecological demand that humans place on the biosphere, which indicates how much pressure is being placed on ecosystems and biodiversity. Despite the fact that the planetary limit for freshwater consumption has not yet been reached, there is a way for estimating one's water footprint and a calculator. The development of additional indicators to calculate the footprint of people on earth will be a topic of continuous research.

The word "pollution prevention" is the widest and hardest to describe specifically. Essentially, it is the act of lessening waste and stopping pollution from leaking into the ground, water, or air. It includes both the waste reduction and the source reduction components. The world population had reached 3 billion people in 1960, and interest in the environment had increased. With a population of over 4 billion people in the 1980s, the idea of preventing pollution gained popularity. When resource conservation and increased efficiency were encouraged for tackling environmental concerns in the 1990s, this environmental approach was regarded as insufficient and was supplemented with the green design concept. The necessity for systems engineering and economic, social, and environmental sustainability considerations grew in the 2000s when the global population reached over 6 billion(Vanek, Albright, & Angenent, 2016). The climate change crises have made carbon footprint analysis a crucial idea and practice. We must define sustainability in order to comprehend the bigger context for carbon footprint measurement because climate change is a main obstacle facing contemporary sustainability efforts. But defining sustainability might be challenging. Within the field of environmental management and its related disciplines, there are almost 300 definitions of sustainability and sustainable development. The number of organizations attempting to define sustainability and sustainable development may actually equal the number of definitions for these concepts. Although there are numerous definitions of sustainability, it is generally agreed that the one outlined by the UN World Commission on Environment and Development (WCED) in the 1987 Brundtland report is the most often used.

2.7 SUSTAINABLE DEVELOPMENT

The Johannesburg Declaration on Sustainable Development, which was the result of the summit, stated that participants agreed that it was everyone's responsibility to advance and strengthen the three interconnected and mutually reinforcing pillars of sustainable development at the local, national, regional, and international levels. The TBL concept is widely used by organizations to assess sustainability performance and results in three categories: economic, environmental, and social. TBL accounting is the process of extending the standard reporting framework to include not only financial performance but also ecological and social performance. The economic bottom line is the economic value produced by the organization for society, to elaborate on each TBL measurement. It differs from the conventional definition of profit in that it takes into account value to the community and society in addition to internal metrics like profit per share and rate of return. Sustainable environmental practices are referred to as the environmental bottom line. This element emphasizes living in peace with nature and preventing harm to the environment in all of its forms. It entails carefully controlling its use of energy and non-renewable resources, lowering production waste, and making waste less harmful before properly and safely disposing of it. The social component of TBL refers to business practices that are ethical and advantageous to the workforce, the community, and the area in which a corporation operates(Brønn & Vrioni, 2001)a practice known as cause-related marketing (CRM. For businesses, this metric may also take into account aspects of giving back through strengthening and developing the local community through initiatives like health care and education. The most challenging to define and potentially problematic and subjective measurement is the social one.

2.8 STAKEHOLDERS

The stakeholders of an organization as opposed to the shareholders are heavily emphasized by the TBL concept. Stakeholders are people, organizations, or groups that the firm or organization's actions have an impact on, either directly or indirectly. The traditional economic indicators of an organization's success are supplemented by the TBL concept with social and environmental considerations. The TBL concept requires evaluating alternatives based on how they will affect the environment in terms of pollution and how those alternatives will successfully integrate into the culture of the organization or community. It goes beyond alternatives solely based on the financial bottom line. For instance, if a business is considering implementing an office recycling program, the management team would need to assess each alternative program on economic factors (initial costs, program costs, revenue, payback period, and rate of return), environmental factors (waste diverted from landfills, energy usage, and changes in emissions), and social factors (employee capability/willingness to participate in the program, and the positive and negative perceptions), as well as environmental and social factors. Although there are many helpful and even some extremely popular definitions of sustainability, there is still no one definition that is recognized by all people and many people would say that there shouldn't be because there are so many diverse perspectives and settings for sustainability. The Brundtland study also states that sustainable development is not a fixed condition of harmony, but rather a process of change. Given this knowledge, it is crucial to comprehend the prerequisites for bringing about sustainable change.

Measuring units in the field of carbon footprint analysis are typically metric tonnes of gas emitted annually for a certain good, activity, or process.A project, strategy, or program should aim to provide quantifiable results that are sustainable and connected to the initial objective. The approval or viability of a project proposal is frequently determined in large part by these outcomes or benefits. These advantages are also the main selling arguments when advocating for GHG and energy reduction among decision-makers and stakeholders. A firm would prefer to establish a scenario in which several advantages can be obtained from a single endeavor. In particular, the company will experience cost savings, the public image will be improved, the environment will be protected, and the organization's stakeholders (including employees) frequently experience a sense of well-being and harmony with the environment as well as the conviction that the organization is defending the greater good for society. The past century has seen a huge transformation in the world. If not already, societies are moving toward a convenience-focused system(Oka, 2021). Waste emissions are rising along with the global population, which has new effects on the environment. For instance, from a practical perspective, emissions rates are rising as a result of the quick industrialization of developing nations, which includes the spread of automobiles. The US produces 17.5 metric tonnes of carbon dioxide per person(Holtz-Eakin & Selden, 1995)a greenhouse gas central to global warming predictions. Estimates derived from global panel data suggest a diminishing marginal propensity to emit (MPE. According to another statistic, the world produces 5.6 metric tonnes of CO_2 equivalent per person, compared to 24.5 metric tonnes from the United States. The impact is made worse by the population's continued growth. According to the U.S. Census Bureau (2011), there are currently 6.6 billion people living in the world, and that number is expected to rise by about 1.2 percent per year. These problems are being made worse by globalization and the growth of developing nations. The environment does not benefit from more people producing more pollution.

Reducing or eliminating an industry's or company's environmental impact is one of the main goals of GHG and energy minimization. When deciding on the course of action for reduction efforts, a grasp of these implications is essential. The contexts in which an organization operates may, regrettably, be separated from these repercussions in both time and space. For instance, a company that is currently releasing GHGs is causing climate change. Although the effects of climate change are currently noticeable, if this trend continues, the repercussions will be considerably more pronounced in the future. As a result, if the effects are anticipated to be worse at a later date than they are today, the company might not be able to absorb its responsibilities. An organization's emissions that contribute to climate change could cause a devastating flood somewhere other than where those emissions originated(Thomas et al., 2019). When it comes to comprehending the two-way connection between an organization's influence on the environment from its GHG emissions and the impact of the resulting climate change on the organization, these indirect consequences can occasionally be a barrier. However, as businesses become more aware of these problems, they will be more motivated to take steps to lessen their carbon footprint.

2.9 CLIMATE CHANGE EFFECTS

Climate change has been connected to a number of common human ailments, including crop failure-related hunger, altered transmission of infectious diseases, and cardiovascular mortality and respiratory problems brought on by heat waves. Due to a lack of long-term, high-quality data sets, the significant influence of socioeconomic factors, changes in immunity, and the emergence of drug resistance, there is still some uncertainty around the relationship between the spread of diseases and climate change. Evidence suggests that the warming trend over the past few decades has already contributed to an increase in morbidity and mortality in many parts of the world, and that climate-health interactions represent growing health hazards under climate change projections. Some regions may be more sensitive than others, such as territory near the Pacific and Indian oceans that has a wide range of rainfall. Clean water, breathing air, and abundant food supply are just a few of the many ways that climate change may have an impact on global health. Extremely hot air directly contributes to mortality from cardiovascular and respiratory conditions. For instance, during the summer 2003 heat wavein Europe, there were over 70,000 additional fatalities reported(Robine et al., 2008).

Ozone and other airborne contaminants are also increased by high temperatures, aggravating cardiovascular and respiratory conditions. According to scientists, urban air pollution results in approximately 1.2 million fatalities annually. Extreme heat also results in greater quantities of pollen and other aeroallergens(D'Amato et al., 2020)biosphere, and biodiversity has become more evident in the recent years. Human activities have increased atmospheric concentrations of carbon dioxide (CO2. Asthma, which affects about 300 million people, can be brought on by these. This strain is anticipated to rise as the temperature continues to rise. Diseases spread by insects, snails, or other cold-blooded animals as well as diseases that are waterborne are all significantly influenced by climatic conditions. Important vector-borne diseases are anticipated to have longer transmission seasons and a different geographic distribution as a result of climate change.

According to the World Health Organization, the area of China where the snail-borne disease schistosomiasis occurs is expected to considerably expand as a result of climate change. In addition, as mosquito populations increase, many eminent experts expect that malaria will spread more widely.

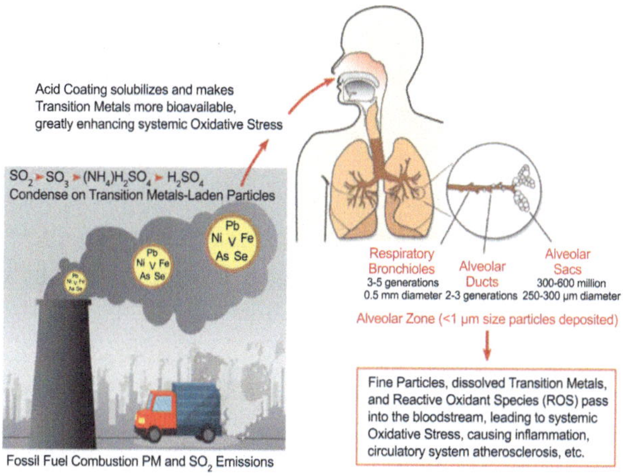

Figure 2.12. Burning fossil fuels. (Source: Maciejczyk P, Chen L-C, Thurston G. The Role of Fossil Fuel Combustion Metals in $PM_{2.5}$ Air Pollution Health Associations. *Atmosphere*. 2021; 12(9):1086. https://doi.org/10.3390/atmos12091086, Creative Commons License)

Burning fossil fuels (Figure 2.12) raises concentrations of various air pollutants like particulate matter, smog, sulfur dioxide, and carbon monoxide in addition to raising global temperatures from the release of GHGs(Rani et al, 2011) (Rani, Singh, Chuhan, Sharma, & Maheshwari, 2011). Air pollution can impair a child's ability to develop their lungs, contribute to the development of early childhood asthma, and cause a variety of respiratory problems in both kids and adults. According to the Santa Barbara County Air Pollution Control District (2011), hazardous air pollutants can have both non-cancer health consequences and raise the likelihood of acquiring cancer. Higher air pollution levels have also been linked to an increased incidence of heart issues, including heart attacks. Reduced GHG emissions and energy use help to safeguard environmental practices and preserve natural resources. The environment is less harmed by fewer mining operations due to energy conservation and the usage of renewable energy sources like wind. Less mining protects natural habitats for species and conserves resources(Watson et al, 2014) (Watson, Dudley, Segan, & Hockings, 2014). The greenhouse effect, which modifies natural ecosystems, can irreversibly impact biodiversity, from microscopic organisms to huge animals. By 2050, a quarter of the world's animal and plant species will be gone, according to several eminent experts. Polar bear population declines due to a lack of food are also being caused by the loss of sea ice in the Arctic. Adelie penguin populations in Antarctica have also reduced as a result of rising temperatures and a loss of sea ice. Since the 1960s, the number of weather-related natural disasters reported on a global scale has more than tripled. Over 60,000 people die as a result of these calamities each year, mostly in underdeveloped nations.

Homes, hospitals, and other vital services will be destroyed by rising sea levels and increased severe weather. According to the World Health Organization (2011), more than half of all people on earth live within 60 kilometers of the ocean. Large populations may be compelled to relocate, which could lead to an increase in communicable diseases and mental disorders. The availability of freshwater is anticipated to be impacted by increasingly unpredictable rainfall patterns. Poor hygiene and an increased risk of diarrheal disease, which claims the lives of 2.2 million people each year (World Health Organization, 2011), might result from a lack of potable water. Extreme water shortages, particularly in developing nations, may result in starvation and drought. The frequency and severity of flooding may both increase due to global warming. Floods can contaminate freshwater supplies, increase the risk of waterborne diseases, and provide habitats for mosquitoes that spread disease, as was discovered in the United States following Hurricane Katrina (Figure 2.13). In many of the poorest areas, the production of staple foods is predicted to decline by as much as 50% in some African countries by 2020 as a result of rising temperatures and unpredictable precipitation(Kotir, 2011).

Malnutrition and undernutrition, which already account for 3.5 million annual fatalities, will become more widespread as a result of this (World Health Organization 2011). The perception of the government agencies' GHG reduction projects' bad economics by American corporations is a prevalent issue. Many business executives think that their organization will lose money if GHG emissions are reduced. This widespread mindset is detrimental and makes efforts to reduce GHG in the United States more difficult. This mindset must be replaced with one that emphasizes how caring about the environment is a smart business decision.

Figure 2.13. Hurricane Katrina. (Source: Englande AJ Jr. Katrina and the Thai Tsunami - Water Quality and Public Health Aspects Mitigation and Research Needs. *International Journal of Environmental Research and Public Health.* 2008; 5(5):384-393. https://doi.org/10.3390/ijerph5050384, Creative Commons License)

2.10 CORPORATE CONCERNS

Corporate environmental concern makes sense since it can be profitable, fosters favorable public opinion, and helps businesses abide by environmental laws. The need for universal waste reduction is highlighted by the fact that there are so many businesses functioning in the United States, each of which produces greenhouse gases (GHG). Leading corporations will work harder to protect the environment and reduce GHGs as a result of the economic case for doing so. When advocating environmentally conscious actions to corporations, the financial advantages of GHG minimization are frequently the main selling reasons. In addition to regulatory compliance, the cost savings from reducing energy use and pollution can transform an environmental decision into a prudent business one that will boost an organization's financial results. When presenting a GHG and energy reduction program to decision-makers of a company, the cost savings produced by the program are frequently its most persuasive benefits. Reducing energy use from non-renewable energy sources that produce GHGs is frequently the one and only major area where businesses may cut carbon emissions. In essence, cutting back on energy use provides the dual advantages of lowering GHG emissions and operational expenses related to energy requirements. Many businesses are shocked to hear that reducing GHG emissions and energy use can be highly profitable(Romm, 2014).

Any organization's profitability depends heavily on its corporate identity and product branding. One of the key responsibilities of any marketing department is to successfully preserve and strengthen these ideals, and environmental initiatives can greatly help to do so. Specifically, a firm can:

- Boost sales by luring environmentally conscientious customers by concentrating on GHG minimization and advertising these efforts,
- Recruiting employees with similar ideals should be improved,
- Attract partners who care about the environment,
- Gain free corporate exposure and, maintain cleaner facilities;
- Increase staff engagement and dedication to the company.

Reducing greenhouse gas emissions also has personal and social advantages. Despite the fact that many of these advantages are somewhat elusive and challenging to quantify, they are important to note since they might serve as selling points when endorsing an environmental initiative. Some of the advantages are: Personal fulfillment from protecting the environment, Long-term environmental sustainability, Cleaner infrastructure, Employee participation (EI), or buy-in at work programs; Healthier settings and a greater standard of life(Kaufman, 2003). Governments are essential in controlling the environment, which includes the air, the land, the water, and all other natural resources. Governments are important institutions for dealing with issues regarding natural resources on a local and international level. Although the economic market has been recognized in recent decades as an appropriate mechanism for regulating environmental quality, markets frequently fail miserably, necessitating government intervention and regulation for the proper, equitable, and sustainable management of the environment.

2.11 EPA

The Environmental Protection Agency (EPA)(Figure 2.14) aims to safeguard and enhance the country's air quality and the stratospheric ozone layer are laid out in the Clean Air Act. The Clean Air Act Amendments of 1990, the final significant revision to the statute, were passed by Congress in that year(Portney, 1990). The Clean Air Act is a federal law in the United States that Congress passed to regulate air pollution on a nationwide scale. To prevent exposure to airborne contaminants that are known to be harmful to human health, it calls for the USEPA to create and enforce rules. The 1963 act had significant revisions in 1970, 1977, and 1990. The 1990 changes to the Clean Air Act included proposals for emissions trading, new clauses for tackling hazardous air pollution, ozone depletion, and acid rain, as well as a national permits program. New specifications for car gasoline reformulation were also introduced by the changes. To classify GHGs as regulated air pollutants, the USEPA has issued regulatory actions under the Clean Air Act. On the its climate change website (http://www.epa.gov/climatechange/initiatives/index.html), some of these are listed. For instance, two programs to lower GHG emissions and increase fuel efficiency were created for transportation sources by the USEPA and the National Highway Traffic Safety Administration (NHTSA), one for heavy-duty trucks and buses and the other for light-duty vehicles. The US EPA published a regulation for stationary sources that specifies GHG emission criteria that specify when industrial facility licenses are necessary for both new and existing facilities.

The Pollution Prevention Act of 1990 established pollution prevention as a national policy and gave the US EPA specific instructions on how to regulate pollutants before they are produced rather than after they are produced(Popp, 2003)which instituted a market for sulfur dioxide (SO2. New reporting requirements were also placed on the industry by the act. The USEPA has created a strategy plan for its pollution prevention program for the years 2010–2014. The Energy Policy Act addresses various aspects of energy production in the US, such as: energy efficiency; renewable energy; oil and gas; coal; nuclear matters and security; vehicles and motor fuels, including ethanol; hydrogen; electricity; energy tax incentives; hydropower and geothermal energy; and climate change technology. For instance, the act offers loan guarantees to organizations that create or apply cutting-edge technology to reduce the generation of GHGs as a byproduct. The amount of biofuel that must be blended with gasoline sold in the US is increased under another provision of the legislation. In the US, an energy bill that would have implemented an emissions trading system was proposed in 2009. By 2050, this law would mandate an 83 percent reduction in the United States' emissions from the reference year of 2005(Spence & Pidgeon, 2009). In response to the Consolidated Appropriations Act of 2008, the USEPA released the Mandatory Reporting of Greenhouse Gases Rule in 2009. The regulation is known as 40 CFR Part 98. The Greenhouse Gas Reporting Program (GHGRP) is the name of the Part 98 implementation program. According to this rule, the US EPA must receive yearly reports from facilities that emit 25,000 metric tonnes or more of GHGs annually.

Figure 2.14. Environmental Protection Agency (EPA).

An international pact associated with the United Nations Framework Convention on Climate Change is the Kyoto Protocol (2011). The main strength of the Kyoto Protocol is that it establishes legally binding goals for lowering GHG emissions, mainly for industrialized nations. Over the five-year period 2008–12, these aims total an average of 5% compared to 1990 estimations. The convention was initially ratified in Kyoto, Japan, in 1997, and it came into effect in 2005. A total of 191 states had signed and approved the treaty as of September 2011. As a sign of their dedication to cutting their carbon footprint, certain items now disclose their carbon footprint. Similarly to this, a quick web search can frequently yield information on the carbon footprints of businesses, cities, states, and governments. The examination of diverse entities' carbon footprints has led to the development of standards, methods, and specifications. The carbon footprint study can be carried out from the standpoint of a product or from the perspective of people, groups, or organizations' actions. The product perspective is in line with the LCA framework and reports the GHG emissions from a good or service's whole life cycle or just a portion of its life cycle phases. The activity perspective, on the other hand, is an annual inventory of the GHG emissions brought on by the actions of specific people, groups, organizations, businesses, or governments(Ostrom, 2012).

2.12 ISO

These four ISO standards apply to all LCA studies; they are not just for assessing items' carbon footprints. Currently, PAS 2050, an international standard for the carbon foot printing of goods and services across their complete life cycle, is the most well-established product carbon footprint standard. Publicly Available Specification is referred to as PAS. The BSI created and released PAS 2050 in 2008, and it was later modified in 2011. By providing requirements, especially for the assessment of GHG emissions within the life cycle of goods and services, PAS 2050 expands on the existing LCA methodologies developed through ISO 14040 and ISO 14044. The application of these criteria in regard to the analysis of the carbon footprint of goods and services is further clarified by these requirements. Currently, certain products are labeled with carbon using the PAS 2050 standard. A company's efforts to lessen its carbon footprint are shown by the Carbon Reduction Label, which is attached to the product. Additionally, it reports the product's GHG emissions per unit of consumption. The purpose of this document is to provide additional specifications and guidance to facilitate the consistent quantification and public reporting of product life cycle GHG inventories. It does this by building on the framework and requirements established in the ISO 14040, ISO 14044 LCA, and PAS 2050 standards, as well as by adding to them. Although only a small number of firms presently employ carbon footprint labels on their products, it is anticipated that as climate change continues to garner more attention, more and more enterprises will pursue carbon labeling. Companies will use these labels and accurately reflect the carbon footprint of their products with the aid of the standards created by the GHG Protocol Initiative, BSI, and ISO(Garcia & Freire, 2014)and ii. Determining the boundaries that the entity has control over, for instance, is a key factor in the inventory. Different standards apply to various sorts of projects because this boundary differs for an individual, a firm, and any other group. There are presently no guidelines for calculating an individual's carbon footprint. There are numerous other tools available for certain sectors, such as those for calculating the carbon footprint of colleges, municipal governments, parks, and other sectors. The corporate standard of the GHG Protocol Initiative is the main manual for conducting company carbon footprint evaluations for enterprises(Hickmann, 2017). The Climate Leaders Program, which began in 2002 and ended in 2011, included several initiatives in its scope. Due to the existence of state-run and non-governmental organization (NGO) climate programs as well as the introduction of the EPA's GHG Reporting Program, the program was discontinued. The GHG inventory is just one stage in a larger carbon management program for groups and organizations. The organization must first make a commitment to lessening its carbon footprint and creating a GHG inventory. Setting targets for emissions reduction follows the release of the inventory. Setting goals results in the creation of a climate action strategy (CAP). Some businesses could think about creating both an energy plan and a CAP. These two strategies differ slightly from one another. An energy strategy where the emphasis may be on lowering energy use would not directly show reduced emissions from reducing process emissions or from moving to other energy sources. Some organizations and stakeholders might not be prepared to address climate change, which would explain why they might be reluctant to conduct a CAP while being highly receptive to creating an energy plan. There isn't yet a direct financial incentive for Americans to reduce their GHG emissions. However, cutting energy use frequently results in lower energy bills. It is quite simple to sketch out the energy plan from the CAP because the construction of the CAP involves energy data and planning.

Although reduction strategies may differ greatly, they should be based on the inventory and workable solutions to cut energy use and emissions in order to fulfill the objective. For instance, a company shouldn't spend too much effort trying to cut emissions that only make up a small portion of its overall carbon footprint. Implementing the action plan and tracking the results are the next steps once an energy plan or CAP has been created. At each stage of the process, communication with all parties involved is important, although it can be most successful after the formation of the plan and then after the review of the progress. The inventory should be completed again, ideally annually, and the process repeated after discussing the progress. Because companies evolve over time, their emissions, target objectives, and reduction strategies may need to be periodically altered in order to accomplish their reduction objectives. This is why the process is iterative or adaptive. The dedication may be institutionalized due to the iterative approach and communication possibilities. All stakeholders will be more committed to the energy and climate goals as long as the work and results are shared(Fenton, Gustafsson, Ivner, & Palm, 2016). Additionally, presenting findings to an audience encourages the CAP team to more carefully examine the data and plans because the team will be motivated to eliminate mistakes and other flaws in their work before presenting it to stakeholders. As a result, improving the quality of the end product from the CAP creation and progress review phases requires effective communication.

2.13 GHG PROTOCOLS

The GHG Protocol Initiative is the biggest and most reputable GHG accounting initiative(Green, 2010). The program began when WRI and WBCSD realized that an international standard for business GHG accounting and reporting would be required in light of changing climate change policies. The GHG Protocol Initiative collaborates with organizations, governments, and environmental organizations from all around the world to create a new breed of programs that are reliable and successful in combating climate change. After the Kyoto Protocol, it is regarded as the second-most crucial climate program for the efficient monitoring and control of climate change. Almost all GHG standards and programs around the world are built on the GHG Protocol Initiative. For various industries or groups, the GHG Protocol Initiative has issued a number of materials. For instance, the Climate Registry, which was formed in 2007 as a 501(c)(3) by U.S. states and Canadian provinces, is the pertinent registry for North America. Senior officials from these organizations oversee the Climate Registry, which was created through a bottom-up partnership amongst North American states, provinces, territories, and Native Sovereign Nations.

The GHG Protocol and the Climate Registryserve as the foundation for the majority of North American GHG inventory instruments. For various objectives, many guiding documents and tools have been created, all of which are based on techniques accepted by the GHG Protocol and occasionally by the Climate Registry. The General Reporting Protocol, Electric Power Sector Protocol, and Oil and Gas Production Protocol are three more sector-specific guidance documents that have been adopted by the Climate Registry. Another illustration is the Climate Leaders Program (from the EPA), a government-industry collaboration that works with businesses to create long-term, all-inclusive climate change policies. The GHG Protocol Initiative has developed five guiding principles for GHG accounting and reporting, particularly in conditions where there is uncertainty. There are numerous gases with GHG characteristics. Some of these gases are not required to be inventoried in accordance with the GHG Protocol's rules even though they have an impact on the climate if they are not covered by the Kyoto Protocol. For instance, anthropogenic GHGs that have been successfully controlled by the Montreal Protocol (Figure 2.15) but are not included in the Kyoto Protocol include fluorinated ethers, CFCs, and perfluoropolyethers(Cousins et al., 2019).

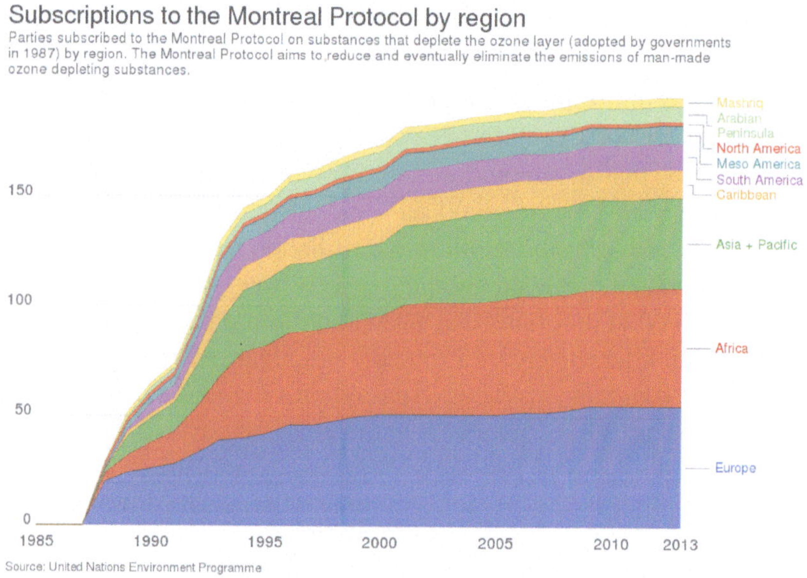

Figure 2.15. Montreal Protocol.

The ozone-depleting gases that are subject to regulation under the Montreal Protocol also have a greenhouse effect. Solvents and disinfectants called fluorinated ethers can also be employed as refrigerant gases. Cleaning agents, dermatological cosmetics, and heat transfer fluids are among uses for perfluoropolyethers. The IPCC did not mention the possibility of global warming for fluorinated ethers and perfluoropolyethers in their 1995 report. According to the GHG Protocol, only six types of gases need to be taken into account. Estimating the effect of GHGs on the climate is one of an inventory's objectives. Each of these gases affects the atmosphere by warming it, but the degree of this warming and, consequently, the final effect, depend on the kind and quantity of the gas. Emissions from all GHGs are normalized to the GWP of CO_2 gas in order to determine one overall impact category(Berners-Lee et al, 2012)& Hewitt, 2012 (Berners-Lee, Hoolohan, Cammack, & Hewitt, 2012)& Hewitt, 2012. For instance, because it is better at retaining heat in the atmosphere than CO_2, CH_4 has a warming effect that is 25 times greater than that of CO_2. As a result, 25 MT of CO_2 are needed to make 1 MT of CH_4. N_2O has a GWP factor of 298 while HFCs, PFCs, and SF_6 often have significantly higher values. The final numbers utilized in the inventory are expressed in units of CO_2 equivalency. The emissions of each gas are computed independently, and they are then converted using GWP values to CO_2 equivalents. GHG inventories commonly utilize GWP factors with a 100-year time frame (Howarth, Santoro, & Ingraffea, 2012). The GWP of a gas, however, depends on the time period investigated, and if the time period analyzed is different, a new GWP factor should be applied. The GWP factors are typically calculated for periods of 20, 100, and 500 years. Each gas has a unique lifetime in the atmosphere as well as a unique ability to absorb infrared light, which can trap heat and lead to global warming. GWP factors for many of the GHGs in the TheSecond Assessment Report (SAR) of the Intergovernmental Panel on Climate Change (IPCC) have been reported.

Any form of institution, including local governments, NGOs, businesses, and their operations, can do a GHG inventory for a project, a business, or for themselves. Their boundaries, both legal and structural, can differ. Whether or not the analysis is being conducted for the public or private sector determines the methodology for defining the organizational boundary. The equity share strategy may be appropriate for the private sector. A corporation accounts for GHG emissions from operations in accordance with its equity stake in the operation under the equity share approach. The control strategy and the equity share approach are options for a corporation. The equity share strategy has no use in the public sector. Therefore, the public sector solely employs the control strategy. Operational control, financial control, or joint control are the three sub approaches that can be applied inside the control approach. The operational control strategy is suggested for local governments. Although there may be some changes in some conditions, the Local Government Operations Protocol indicates that in most cases the local government inventory results will be similar regardless of whether operational or financial control is employed. In either case, the organizational boundary strategy chosen should be applied consistently rather than alternating between the two strategies. It is possible to separately count an organization's direct and indirect emissions by separating the emissions in this manner. Additionally, the indirect emissions of one business will be the direct emissions of another. Direct emissions may come from fixed combustion that generates power, heat, or steam, as well as from mobility sources like cars.

CHAPTER 3

Sequestration of Carbon Dioxide

CONTENTS

3.1 INTRODUCTION

BTEX chemicals are the collective name for the isomers of benzene, toluene, ethyl benzene, and xylene (Aranda et al., 2010) (Figure 3.1). The hydrocarbon sector is particularly interested in the containment and disposal of these chemicals because it is known that they are hazardous to people.

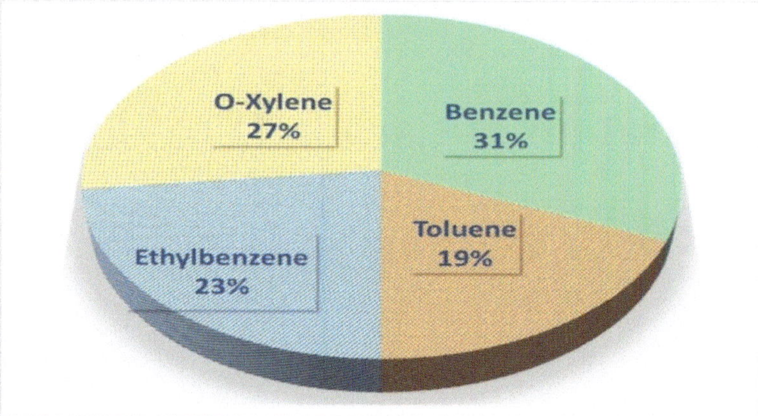

Figure 3.1. BTEX chemicals.

Accidental spills or leaks from underground gasoline storage tanks are frequently linked to BTEX environmental contamination. Because natural gas and other produced fluids contain trace levels of BTEX compounds, regulations governing clean air emission standards directly affect the natural gas processing business. Aqueous solutions of organic molecules with a high affinity for the BTEX components are frequently utilized as liquid solvents to remove H_2S and CO_2 from gas streams. A complicated phenomenon involving several interconnected process variables, such as operating pressures and temperatures, amine composition, amine circulation rates, and others, determines how the BTEX compounds are distributed throughout the many streams of a natural gas processing facility. The quantity of BTEX compounds that wind up in the acid gas product leaving the amine regenerator is of special interest in acid gas injection. If not taken into consideration during the design stage, the presence of trace amounts of BTEX compounds in the acid gas may result in the unexpected production of a non-aqueous liquid phase in the compressor.

There is a need for precise vapor-liquid equilibria (VLE) and/or vapor-liquid-liquid equilibria (VLLE) data for BTEX and related hydrocarbons in amine treatment solutions under rich amine conditions in order to calculate the quantities of BTEX chemicals that will be present in the gas(González et al., 2018)the recovery of the extracted hydrocarbons has been much less investigated, being a required task to complete the global separation process. Taking into account the negligible vapor pressure of the ILs, this step could be easily carried out by flash distillation, which requires the study of vapor-liquid equilibrium (VLE. Data from operating plants can also be used to validate the results of any thermodynamic model. In order to demonstrate general trends in amine plant BTEX absorption using computer simulation, researchers collated the experimental data available for hydrocarbon and BTEX solubility in amine treatment solutions. The hydrocarbon components benzene, toluene, ethyl benzene, and dimethyl benzenes were examined (xylenes) in many studies. Initial calculations showed that each of the three xylene isomers had a similar effect on the phase behavior of the acid gas combinations. An experiment was done utilizing pure H_2S to conduct a sensitivity evaluation and determine a tolerable range of BTEX compositions. The model was applied to various benzene and *o*-xylene compositions ranging from 0 to 5000 ppmv to calculate the dew point temperature at 4000 kPa. Aromatic chemicals do not significantly affect dew point temperature below concentrations of 100 ppmv, increasing it by less than 1°C. When BTEX compounds are present as gases, the phase boundary widens at a given pressure; this effect is more prominent in acid gases with higher CO_2 content.

When BTEX compounds are present in gas, the dew point temperature rises at a given pressure; this effect is more noticeable in gases with higher H_2S content. This is partly because high H_2S combinations cause the gas phase envelope to shift to higher temperatures. BTEX compounds increase the dew point temperature in the following order: benzene, toluene, ethyl benzene, and *o*-xylene, with *o*-xylene having the most pronounced effect, at equal concentration in the acid gas and at equal pressure. If BTEX compounds are present in the acid gas at levels less than 100 ppmv, the acid gas dew point is relatively unaffected. The PVT (Pressure-Volume-Temperature) unit is typically used to carry out the standard fluid phase behavior test. The exploration of near-critical gas condensate/volatile oil reservoirs and gas-injected improved oil recovery methods are just a few high-pressure technologies where it is a major concern. Both domestically and internationally, CO_2 displacement technology is acknowledged as a key and well-established tool for oil and gas improved recovery. Miscible gas injection is regarded as an effective enhanced oil recovery technique that could lessen the ability of capillary forcesto trap fluid. Due to the complexity of multi-component reservoir fluid, despite the fact that certain PVT fluid phase behavior data are accessible in the publications, they are still insufficient. The HP6890 gas chromatograph was used to analyze the gas phase. Using the HP5890A to simulate the distillation process, the liquid phase was examined(Sun & Chen, 2005). The reservoir fluid composition was then formed by mixing the compositions of the gas and liquid phases using the gas oil ratio (GOR). The vapor pressure osmometer (VPO) was used to calculate the molecular weights of the oil phase; the resultant molecular weight was 420 g/mol.

3.2 OIL SYSTEMS

The link between the volume and pressure of the reservoir oil was constructed using the PVT data for various CO_2 injected components. The collected PVT data might then be used to compute the density and boiling point pressure of the reservoir fluid at various pressures, which is helpful for calculating phase behavior features like the solubility of injected CO_2 in oil. There is no discernible difference between the gas phase and liquid phase when the CO_2 concentration reaches 65.0 mol percent, indicating that the presence of first contact miscibility may be present. However, the boiling point pressure value increases more dramatically with an increase in CO_2 molar concentrations when the CO_2 content is higher than 45 mol percent. The appropriate CO_2 injection concentration is also chosen using the boiling point pressure data. The amount of CO_2 injected changed from 0 to 65.0 molpercent, which significantly reduced the viscosity of the reservoir oil. When 65 molpercentCO_2 was injected, the viscosity value might drop from 10.6 cP to 1.1 cP at roughly 30 MPa. The reservoir oil viscosity is observed to increase with pressure when the experimental pressure is higher than the saturated value; conversely, when it is lower than the saturated pressure, the reservoir oil viscosity increases with pressure drop. The viscosity of the residual oil increased as a result of more CO_2 being released from the reservoir oil due to the pressure drop. However, the reservoir oil viscosity in single phase conditions did not considerably reduce with the subsequent increase of CO_2 injecting concentration after CO_2 content was higher than 44.7 mol percent(Abedini & Torabi, 2014)the solubility of the CO2 in the light crude oil and oil swelling factor due to the CO2 dissolution in the oil phase were determined at T = 30 °C and various equilibrium pressures ranging from atmospheric pressure to Peq = 12.55 MPa. Later, series of immiscible and miscible CO2 huff-and-puff tests were designed and carried out at various operating pressures (i.e., Pop = 5.38–10.34 MPa. The boiling point pressure only rises from 11.28 MPa to 14.14 MPa as the CO_2 injecting concentration increases from 0 to 44.7 mol percent. However, the boiling point pressure rises from 14.14 MPa to 25.0 MPa as the CO_2 injected concentration rises from 44.7 mol percent to 65.0 mole percent. A high CO_2 concentration is not necessary because there is an appropriate CO_2 injecting concentration due to the decrease in viscosity and boiling point pressure. After the water displacement procedure, a significant amount of reservoir water is present in the stratum. Since capillary forces heavily influence reservoir fluid movement, proper interfacial tension calculation is very important. The concentration of CO_2 that was injected also has a significant impact on the interfacial phenomena. There is no discernible difference between the gas phase and liquid phase when the CO_2 concentration reaches 65.0 mole percent, indicating the possibility of first contact miscibility under the corresponding CO_2 injected content. Evidently, when the CO_2 level increased, the viscosity of the crude oil for CO_2 injection reduced. The CO_2 injection is important for reducing the viscosity of the crude oil in the Jilin reservoir.

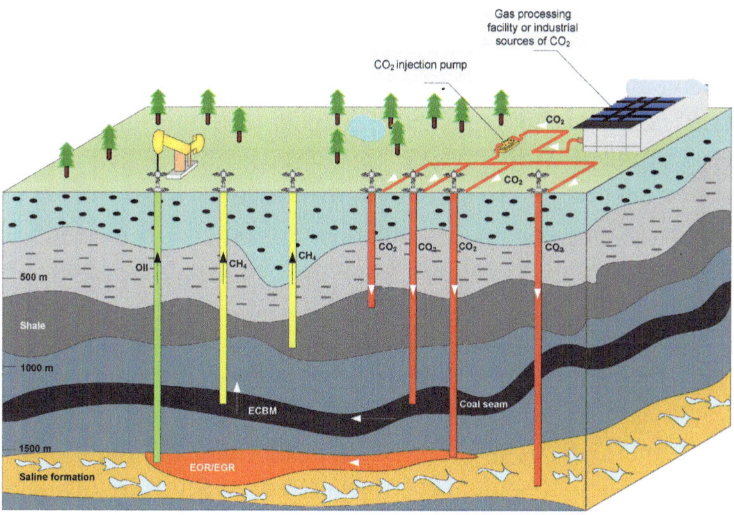

Figure 3.2. CO_2-injected oil system.

However, the reservoir oil viscosity under single phase conditions did not considerably reduce with the subsequent rise of CO_2 injecting concentration after CO_2 content was higher than 44.7 mol percent. When the CO_2 content fluctuates from 0 to 65.0 mole percent, the interfacial tension for CO_2 injected oil/reservoir water appears to decrease with an increase in CO_2 injected molar concentration(Khaksar et al, 2016)the carbonated brine injection causes the trapped oil to lose viscosity and increase its mobility. The current study is about smart water and carbonated smart water injection and sets out to investigate the effects of soluble ions on reducing the interfacial tension (IFT (Khaksar Manshad, Olad, Taghipour, Nowrouzi, & Mohammadi, 2016)the carbonated brine injection causes the trapped oil to lose viscosity and increase its mobility. The current study is about smart water and carbonated smart water injection and sets out to investigate the effects of soluble ions on reducing the interfacial tension (IFT).

The CO_2-injected oil system (Figure 3.2) neared perfect miscibility at a CO_2 composition of 65.0 molpercent, while the interfacial tension data of CO_2-injected crude oil/reservoir water altered somewhat with pressure. The density and viscosity characteristics of the fluid must be thoroughly understood in order to design an acid gas injection (AGI), or CO_2 injection scheme. The viscosity is needed to evaluate frictional pressure losses from pre-compression to the reservoir, while the density is needed to compute pressure gains from static head. In order to reconcile differences within existing data sets and further parameterize models, it is desirable that the literature data be expanded to cover the applicable temperature and pressure zones, particularly at conditions for AGI methods. Although pure CO_2 and methane have undergone extensive research into their density and viscosity characteristics, little is known about H_2S under conditions that are relevant to industry, particularly with regard to their viscosities at higher pressures. We took into account that gas streams flowing through conventional compression cycles involve a large range of temperatures from T = 0 to 150°C in order to identify the range of conditions that would be suitable for industry. Even higher goal reservoir pressures (p > 700 MPa) can be anticipated if the fluid is being employed to maintain reservoir pressure. Thus, T = 0 to 150°C and p = 0.1 to 100 MPa are the general temperature and pressure ranges that research is focused on. We can analyze a number of pure acid gas components and acid gas mixtures within this range of conditions, starting with CO_2(calibrant), CH_4, and H_2S. Because of the fluids involved, the wide temperature and pressure ranges described are experimentally difficult.

3.3 DENSITY AND VISCOSITY MEASUREMENTS

Densities under high pressure can be precisely measured using a variety of techniques. Isochoric vessels with accurate pressure measurement and stable temperature control are simple, provided the mass of the fluid can be determined with high confidence. The vessel can be heated to the necessary temperatures and the resulting pressure can be measured using the isochoric method. For controlled isothermal fluid expansion, a second vessel may be employed. The benefit of vibrating tube densitometers, or VTDs, is that they can measure densities for static or flowing fluids and are applicable over a wide range of densities, typically p = 1 to 2000 kg m^3. The brewing and distillation industries have historically utilized VTDs to calculate alcohol content. Researchers can produce reliable findings up to very high pressures as long as they are combined with adequate temperature and pressure management. Capillary viscometry, oscillating viscometers, and other methods are among the various high pressure viscosity techniques that could be used. Most viscometers are developed to measure liquid or gaseous viscosity and are rarely designed for a wide range of viscosity, unlike some of the densimeter approaches.

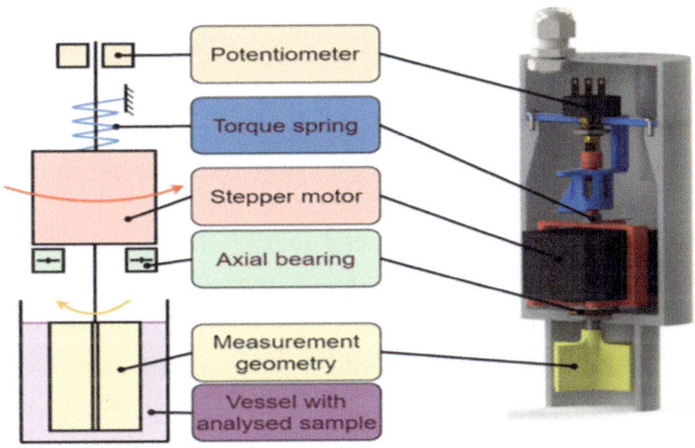

Figure 3.3. Viscometer prototype for remote in situ analysis of sludge.

They all, in general, entail some item falling through a static fluid while being pulled backward by an opposing drag. The precision of the velocity measurement, or the measurement of the travel time for the object to move a predetermined distance, determines the accuracy of the viscometer. The falling object's density (buoyancy), shape (drag and tolerance), or additional friction can all be provided by allowing the object to roll or slide across the surface of an inclined tube in order to reduce the elapsed time. Many methods, including optical and electromagnetic techniques, have been applied to improve elapsed time measurement. Because the fluid is static, falling body viscometers are well suited for high-pressure applications; nonetheless, they are frequently employed for liquid phase conditions as opposed to gas phase situations, where the viscosity is very low. Aside from vibrating wire and vibrating piezoelectric quartz crystal, other high pressure instruments that can be used to test low viscosity include oscillating discs, vibrating wire, and oscillating discs. Cambridge Viscosity Inc. has marketed a variation of the falling cylinder viscometer. This kind of viscometer, referred to as a Cambridge Viscometer, was created for high viscosity fluids, but it has proven possible to lower the minimum observable viscosity by employing a hollow cylinder with less tolerance between the cylinder and vessel wall. The main benefit of this viscometer is its small size and wide range of viscosities; for instance, more cylinders may be acquired to hold viscous fluids.

Two magnetic coils located outside the stainless steel chamber are used to measure the force required to move the piston a predetermined distance (0.2”) at a steady speed. The travel time is monitored and converted into absolute viscosity by switching the power to the coils. As long as the tests were conducted at low pressure, the parameters established were effective for the various fluids evaluated during this operation. The tolerance between the piston and the viscometer chamber also increases with large changes in temperature or pressure, reducing the resistance to motion. Hydrogen sulfide, carbon dioxide, water, and trace amounts of light hydrocarbons make up the acid gas stream produced by the gas treatment process. For storage and disposal in many field applications, acid gases are compressed and re-injected into deep salt aquifers or depleted oil reservoirs.

The viscosity of acid gases and single phase mixtures of hydrocarbons, water, and acid gases is the topic of this work, which also provides a tool to predict their viscosity. Scientists created the expanded fluid viscosity correlation based on their discovery that fluidity, which is the inverse of viscosity, increases when a fluid expands. There is a defined compressed condition of a fluid where the molecules are too closely packed to flow, at which point the viscosity reaches infinity. The fluid’s expansion from this stage was then correlated with the viscosity. The experimental viscosity and density data that served as the foundation for the correlation’s framework were examined for both pure hydrocarbons and their mixes. Although any density model might be used for this purpose, it should be noted that the correlation parameters rely on the density model selected. The correlation must be modified for commonly found non-hydrocarbons like water, hydrogen sulfide, and carbon dioxide in order to be used for acid gas injection applications.

3.4 VISCOSITY DATA AND MEASUREMENTS

Experimental mixture viscosity data for these systems are needed to investigate the precision of the correlation predictions for acid gas mixtures. The correlation follows the change in phase conditions from a diluted gas to a dense fluid smoothly. Keep in mind that all information for the H_2S-containing binary and ternary mixtures comes from molecular dynamics (MD) simulations(Brechet, 2004). By comparing the correlation predictions made by various intermolecular potential fields and MD simulations to the mean of the anticipated viscosity values, the average absolute relative deviation (AARD) and maximum absolute relative deviation (MARD)for these mixes were computed. Water content can have a sizable impact on the viscosity of gas mixtures for acid gas applications. The ability to forecast the viscosity of the acid gas injected below and dissolved in brine is useful for reservoir engineers. In order to investigate the accuracy of the model predictions and enhance its performance, experimental data on the viscosity of these systems are needed. Water, carbon dioxide, and hydrogen sulfide are examples of non-hydrocarbons that were added to the expanded fluid based viscosity model that was originally designed for hydrocarbons. In order to take into account hydrogen bonding, a temperature dependency term was added to the compressed state density. With this change, the correlation had an overall AARD of less than 4% and suited the viscosity data for the pure components. The correlation indicated that combinations of light hydrocarbons and acid gases would have a viscosity with an overall AARD of less than 7%.

Without any modifications or additional parameters, the conventional mixing rules for the correlation were applied. A reliable technique for the prediction of the hydrocarbons and non-hydrocarbons found in acid gas injection applications is the extended fluid-based viscosity model and associated mixing rules. It is a useful tool for simulators due to its applicability for both solid and liquid phases and quick computer code execution. The thermodynamic engine is prepared to carry out physical property calculations of the described system after components are connected to the package. Nearly all of the NH_3 is transformed into ammonium ions at this minimal threshold, and the ions from the gases are saturated. This behavior of NH_3 solution can be employed as a controllable operational variable for gas treating procedures as any more gases introduced would proceed into the vapor phase from this point on, resulting in a sudden increase in total pressure. Based on this, it would be advantageous to develop an NH_3 recycling process that could absorb CO_2 at low temperatures. An engineer can use this behavior to choose the best process settings to remove gases since the impact of NH_3 concentration on CO_2 absorption is more pronounced at high temperatures. The total vapor pressure drops first and reaches its lowest point when the molality of gases rises. Nearly all of the NH_3 is transformed into ammonium ions at this minimal threshold, and the ions from the gases are saturated. This behavior of NH_3solution can be employed as a controllable operational variable for gas treating procedures as any more gases introduced would proceed into the vapor phase from this point on, resulting in a sudden increase in total pressure.

A PVT vessel with a capacity of around 135 ml, an automated pump, a gas chromatograph, a dry gasometer, an air bath with a constant temperature, a flash separator, and a ground separator made up the experimental equipment. A HP-6890 Gas Chromatograph was utilized for the analysis of the used gas samples. The heat loss to the surroundings of a pipeline buried in the ground can be easily confirmed by noting the difference in fluid temperature between the pipe's input and outflow. This could also be done in a well, but installing temperature recorders is expensive, even temporarily. Furthermore, operators are discouraged from taking such temperature measurements due to the possibility of a wireline failure. In comparison to the conditions of the pipeline, the physical conditions in the tubing and the nearby rock are somewhat different. Digging the pipe trench revealed the type of soil that the pipe is buried in.

The well hole is several millimeters larger when it is drilled than the casing that will be put in. Typically, some parts of the well bore are washed away, leading to heavier concrete sections there. No well is perfectly vertical, therefore even when the casing is anchored in the hole, it may not be exactly in the middle of the well bore. Additionally, the tubing is installed and after the well is finished. The tubing is compressed at the lower part of the well, which means that it is not in the center of the casing but rather is in direct contact with it as a spiral column.

The tube is in tension in the upper part, but because the well is not really vertical, it still touches the casing. The tubing joints' couplings have a little bigger outside diameter than the pipe itself, which prevents it from directly touching the casing. Water or a hydrocarbon liquid, such as diesel fuel or stabilized condensate, is used to fill the area between the tube and the casing. There are significant differences in these two types of fluids' thermal conductivities. The temperature difference between the tubing wall and the casing wall will also cause convection eddies. The heat must pass through the tubing wall, the casing, the concrete sheath enclosing the casing, and finally to or from the rock for there to be heat transfer between the acid gas in the tubing and the surrounding rock.

3.5 THERMAL CONDUCTIVITY

The thermal conductivities of different rock formations vary(Fuchs & Balling, 2016)detailed knowledge of rock thermal properties are paramount for a reliable parameterization of layer properties and boundary conditions. Despite the state-of-the-art in other research fields (e.g. hydrogeological ground-water models. Because there are too many unknowns, calculating the overall heat transfer coefficient in light of the peculiar placement of the tubing in the casing (hole), the various concrete sheath thicknesses, and the various thermal conductivities of rock formations, both porous and impermeable, would produce uncertain results. In light of the fact that substantial variations create relatively modest differences in the compressor discharge pressure, especially at higher speeds, the estimation of an overall heat transfer coefficient comparable to the relationship for pipeline heat transfer is a plausible strategy. The coefficient should be changed to match the measured temperature after determining the temperature using an instrument. The injectivity index and the coefficient of friction in gas flow in pipes are both determined by the transport property known as viscosity. An acid gas injection compressor's discharge pressure requirement can be calculated using a computer model. The injection capacity of the injection zone has a significant impact on the pressure, which is mostly determined by the reservoir pressure. This situation raises questions about how accurately equilibrium water contents of high- and low-pressure gases containing extremely high quantities of CO_2 and/or H_2S can be calculated.

The water content of the treated gas and the solubility of hydrocarbon, acid gas, and particularly the BTEX in the water-rich glycol are two crucial components of thermodynamic modeling. The acid gases and hydrocarbons, particularly the BTEX components, are additional components whose solubility in TEG is important. The noticeable heating produced by water absorption is equal to its latent heat of condensation. It should be noted that high TEG viscosity is a consideration and will negatively impair internal performance when compared. The lowest temperature has a viscosity of roughly 19 cP. At the solvent moisture levels found in these simulations, viscosity is not greatly influenced by water content. The wet gas continues to heat the condensate as it trickles through the bed, but during the first 18 inches, it reaches a peak temperature of 194.5°F before abruptly dropping to roughly 160°F. As one moves down the column, the gas gets hotter and eventually rises quickly to 294°F. The solution for the gas phase is quite simple: feed that has been warmed to 300°F enters the column at the 2-ft level and evaporates into its liquid and vapor components. The hot feed merely collides with a cold reflux water stream at the feed site, causing a dramatic change in vapor temperature. The water in the reflux, however, cools so considerably.

When the vapor rises above the liquid in this portion, its water content is very close to being in equilibrium with the liquid. There is more than 95 weight percent TEG in the liquid and just 5 weight percent (17–18 mole percent) waterits use in the air conditioning market is still very limited due to its high installation cost, intensive carryover, and high cost of operation and maintenance associated to the corrosion of the current desiccant liquids. Substitutes for the traditional desiccant liquids with better properties and no corrosion are highly needed. Ionic liquids, which are salts comprised of organic cations and inorganic anions or organic anions, have high thermal stability, negligible or no vapor pressure, varied solubility in water, low or no corrosion to metals, and low driving temperatures to achieve dew point temperatures. All of those characteristics make them as perfect substitutes for traditional desiccant liquids. Up to date, a very few research groups investigated ILs as potential alternatives to traditional desiccant liquids for LDAC. The study in the paper aimed to identify the ionic liquids as ideal desiccant liquids to achieve better cost-effectiveness and higher system performance of LDAC. There were 13 different ionic liquids (IL. The mass transfer rate is not a restriction on the evaporation process. The pace at which the latent heat of vaporization required by the humidification process can be transferred from the bulk liquid to the gas-liquid interface is what regulates the rate of humidification. The rate of humidification just above the feed point is really controlled by conductive heat transfer over the liquid layer covering the packing, indicating that this process is one of heat transfer. Heat transfer rate is a constraint on the process. As a result of the mass transfer driving force for humidification, the reflux water is cooled by about 33°F since enough heat is removed from it. The liquid meets hot vapor from the reboiler. This causes the liquid to become warmer as the vapor is cooled, and some of the heat transfer is caused by water that is transferring from the vapor back into the liquid towards the bottom of the section. This modification made it possible for the glycol purity to rise from 99.79 weight percent to 99.96 weight percent TEG. The gas dew point should be significantly lower than the temperature required for hydrate formation at this water content. Therefore, it would seem that gas stripping is a more efficient method of enhancing dehydration system performance. Acid gas injection calls for compression of the gas to extremely high pressure. If the gas is moist, compressing it will cause liquid water to escape; this liquid will then be under intense pressure of an acid gas. It is therefore interesting to compare the dehydration of pure but water-saturated H_2S with that of an equivalent amount of gas.

3.6 AMINES FOR REMOVING CO_2

For more than 60 years, amines have been employed to remove CO_2 from gas streams in many sectors of the oil and gas industry (Olajire, 2010) large decreases in CO2 emissions through capturing and separation will be required. This article reviews the possible CO2 capture and separation technologies for end-of-pipe applications. The three main CO2 capture technologies discussed include post-combustion, pre-combustion and oxyfuel combustion techniques. Various separation techniques, such as chemical absorption, physical absorption, physical adsorption, cryogenics, membrane technology, membranes in conjunction with chemical absorption and chemical-looping combustion (CLC. Amines are mostly employed in refining to remove H_2S from hydrocarbon gas (and liquid) streams from a variety of different sources. The gas is finally burned when the amine mechanism removes the majority, if not all, of the carbon dioxide from it. Heat stable salts, which are present when stronger acids than H_2S or CO_2 are present and they create a very strong bond with the amine, are a common source of issues for refineries. Amines are additionally employed globally in natural gas processing facilities to eliminate H_2S and/or CO_2. Many different types of amines are used depending on the situation because the gas streams being treated in these plants vary substantially in pressure and composition. In the past, oxygen entering the amine system and causing solution degradation has been a problem for gas plants. As an alternative to further lowering the H_2S level of the gas leaving a sulfur plant, tail gas treating is done in both gas plants and refineries. A tail gas treatment facility's main objective is to keep CO_2 in the gas while removing H_2S from it selectively. Because of the incredibly low pressures used, it is quite challenging to separate the acid gas from the gas.

It is conceivable to use amines in the carbon capture sector, however, there are certain special challenges. Carbon dioxide is removed from a very low pressure, highly oxygenated gas stream during carbon capture(Allam et al., 2017)oxy-fuel power cycle that utilizes hydrocarbon fuels while inherently capturing approximately 100% of atmospheric emissions, including nearly all CO2 emissions at a cost of electricity that is highly competitive with the best available energy production systems that do not employ CO2 capture. The proprietary system achieves these results through a semi-closed-loop, high-pressure, low-pressure-ratio recuperated Brayton cycle that uses supercritical CO2 as the working fluid, dramatically reducing energy losses compared to steam- and air-based cycles. In conventional cycles, the separation and removal of low concentration combustion derived impurities such as CO2 results in a large additional capital cost and increased parasitic power. As a result, removal in conventional cycles can increase the cost of electricity by 50% to 70% [1]. The compelling economics of the Allam Cycle are driven by high target efficiencies, 59% net for natural gas and 51% net for coal (LHV basis. When employing amines, the creation of heat stable salts, chemical breakdown, and low pressure treatment are problems in the carbon capture process. Amines are medium strength bases that are utilized to remove CO_2 from gas streams and have a pH of around 10. When water is present, the acidic CO_2 interacts with the amine to create a salt. Rich amine is an amine that has interacted with carbon dioxide. Primary or secondary amines react with CO_2. The gas and amine come into contact with one another. At the bottom of the tower, gas enters, and at the top, amine. CO_2 steadily moves from the gas into the liquid as the two flow against one another. When the gas finally comes into contact with amine, it is at the top of the tower. The amount of CO_2 in the gas will be equal to the amount of CO_2 in the amine; the less CO_2 in the amine, the easier it will be for the CO_2 to escape from the gas. The processed gas that emerges from the absorber will primarily be made up of hydrogen and water and can either be vented to the atmosphere or burned. Before entering the regenerator tower, the CO_2-loaded rich amine is heated in a heat exchanger. The regeneratorbreaks the bonding between amine and CO_2. As the amine descends through the tower, it is heated to reverse the process. Steam is provided as a source of heat, and it is produced at the reboiler at the base of the tower.

The reboiler is the amine plant's biggest energy consumption, making it the main subject of studies on plant optimization. A heat source powers the reboiler; this source is frequently low pressure steam, but it can also be hot oil, glycol, or even directly burned. The amine solution water component boils within the reboiler, creating steam. The heat from the steam is then transferred to the endothermic process that destroys the bond between the amine and CO_2 as it rises through the regenerator tower. The reboiler's steam is used for three major things: Steam must be escaping the top of the regenerator in order to sweep the newly liberated CO_2 out of the tower and to serve as a source of reflux flow for the system. Sensible heat is required to raise the temperature of the amine to its boiling point. Reaction heat is required to break the bond between the amine and CO_2. After cooling in the exchanger and the amine cooler, the regenerated amine exits the reboiler. It is filtered and is prepared to be injected again into the absorber. Nearly all of the gas stream leaving the regenerator is CO_2. It can be compressed and pumped underground for long-term/permanent storage or sequestration, or it can be liquefied and sold. The physical dissolution of CO_2 into the water phase initiates the reactions of CO_2 in the gas phase with an aqueous amine solution. Any significant reaction requires the transfer of the CO_2 molecule from the gas phase to the liquid phase.

Although some interfacial interactions are feasible, liquid phase reactions predominate.In contrast to nucleophilic reactions, which are often diffusion-controlled, acid-base reactions are quick. However, the acid-base pathway for the removal of CO2 is slow since slow-generating carbonic acid must be produced first(Spooner & Engel, 2011). The transfer of CO2 from the gas phase to the liquid phase is the initial event in a CO2 reaction with amine. The CO2 molecule is then hydrolyzed, resulting in the formation of carbonic acid and bicarbonate. The core carbon of CO2 is directly attacked by the amine in this reaction phase. The intermediate then experiences an internal H transfer, which results in the production of a carbamic acid. The amine and this continue to combine, producing the ammonium salt. Since there is no H in the case of a tertiary amine, there is no reaction. It is crucial to note that as the amine transitions from a primary to a secondary amine, the reaction slows down and becomes less effective. This is so that the amine N atom cannot physically access the CO2 due to the R groups.

3.7 ACID-BASE REACTION

Despite the fact that the end result of an acid-base reaction is a salt, the product is instantly dissolved in the aqueous solvent and forms a group of ions. These ions have a thick layer of water molecules covering them, making them extremely hydrated and preventing reaction reversal under typical conditions. An acid-base reaction is an alternative mechanism. This reaction happens with the byproduct of the interaction between CO_2 and water rather than with CO_2 itself (carbonic acid). The alkanol R substituents in an amine have little effect on this quick reaction. Depending on the inductive effect of the substituents, some amines will have faster reaction rates. Tertiary amines react with acids and bases more quickly than secondary and primary amines. These two types of reactions (acid-base and nucleophilic) show how CO_2 is taken out of a stream using amine solutions. MEA is one of the best amines for removing CO_2 from a gas stream(Bougie & Iliuta, 2012)oil, and gas industries, as well as in environmental protection. The choice of a certain amine (single or blended amine. This specific molecule has good acid-base reactivity and is the most reactive via a nucleophilic pathway. Alkanolamines come in varieties and are all capable of removing carbon dioxide from gas streams to varying degrees. However, only a few amines are suitable for this procedure due to the special characteristics of post-combustion gas from a power plant.

3.8 MONOETHANOLAMINE (MEA)

A primary amine called monoethanolamine (MEA) is the most effective and trustworthy CO_2 elimination product on the market. It is inexpensive and easily accessible compared to other amine kinds. MEA offers a number of benefits and drawbacks. Since MEA is the most reactive amine, it forms the strongest bond with carbon dioxide. In contrast to other amines, it is very dependable and will continue to remove CO_2 under more adverse working conditions or larger levels of pollution. The molecular weight of MEA is the smallest of any amine, and it is not a particularly complex molecule. Because of this, even a weak MEA solution is extremely effective. It is preferable to run MEA at low strengths while still being able to remove CO_2, as this reduces the danger of corrosion caused by deteriorated MEA and heat stable salts. Of all the amines, MEA is the most affordable and widely accessible. In contrast to secondary or tertiary amines, MEA can be recovered at atmospheric pressure. By heating stable salts that would otherwise accumulate in the solution, atmospheric recovery removes degradation byproducts that would otherwise increase the solution's propensity for corrosion and weaken the link between amine and CO_2. The tight link between MEA and CO is advantageous in the absorber but disadvantageous in the regenerator since it takes a significant amount of energy to break the MEA:CO_2 bond. Further, MDEA, a tertiary amine, has historically been utilized in conditions when CO_2 elimination is not desirable. The CO_2 molecule and MDEA do not directly interact. However, with the addition of a promoter (often piperazine), MDEA may remove

CO_2 without the two drawbacks of MEA: degradation and high reaction heat(Idem et al., 2006)lean and rich loadings, CO2 production, and solvent stability were performed by comparing the performance of aqueous 5 kmol/m3 MEA with that of an aqueous 4:1 molar ratio MEA/MDEA blend of 5 kmol/m3 total amine concentration as a function of the operating time. The tests were performed using two pilot CO2 capture plants of the International Test Centre for CO2 Capture (ITC. Although it relies partly on the desired amount of CO_2 in the treated gas, it can be challenging to successfully do this under low-pressure settings because partial pressure is the driving force to transfer CO_2 from the gas phase into the liquid phase. Lower CO_2 loaded amine is needed for deeper levels of CO_2 elimination. The amine will react with the oxygen in the input gas to produce degradation products and salts that can withstand heat. Unfortunately, in the presence of oxygen, all amines decompose. Degradation is also brought on by CO_2, but only in primary and secondary amines (MEA and DEA). The viscosity of an amine can increase, making it more difficult to heat and cool, and the breakdown products and heat stable salts may make the solution more corrosive.

The major method for preventing or minimizing degradation is to keep the absorber's temperature as low as possible. A reaction between amine and CO_2 is exothermic, which means heat is produced. Interstage coolers, which remove the amine from the center of the absorber, cool it, and then re-inject it, are used to lower the temperature. Depending on the tower's height, there can be more than one cooler. On MEA systems, thermal reclaimers are utilized to eliminate degradation byproducts and heat stable salts. While operating at a higher temperature, the reclaimer resembles the reboiler in appearance. The higher boiling point breakdown products are left behind but the amine and water are evaporated and reintroduced to the regenerator. Because the boiling points of the amine and the degradation products are too near to one another, MDEA cannot be recovered at atmospheric pressure. This is a simple solution to the accumulation of salt products that degrade and become heat stable, but it should only be used as a last resort. The expense of disposal of the current is in addition to the cost of purchasing fresh amine. Because heat is required to stop the reaction between amine and acid gas, the reboiler consumes the majority of energy in amine systems. The reboiler is necessary to generate sensible heat as well as reflux heat, in addition to reaction heat.

CHAPTER 4

Carbon and Sustainability

CONTENTS

4.1 INTRODUCTION

The average atmospheric temperature has increased as a result of the greenhouse gas, carbon dioxide (CO_2) emissions(Letcher, 2020). Despite a pandemic-related slowdown in the global economy as of May 2021, atmospheric CO_2 levels are at 419 ppm.

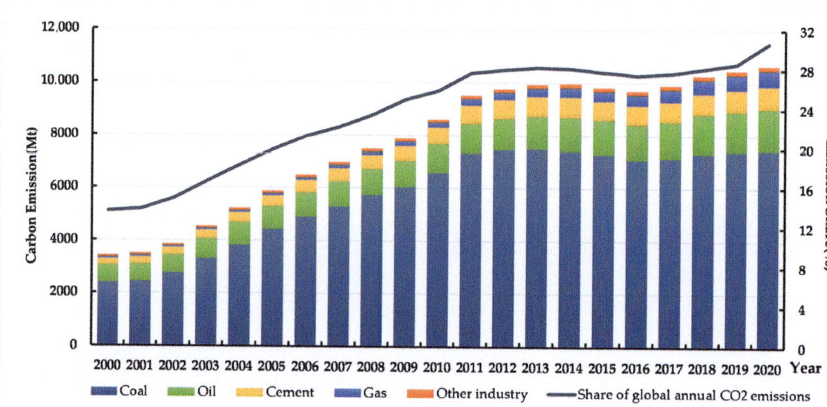

Figure 4.1. Global CO_2 emissions.

Global CO_2 emissions (Figure 4.1) must be lowered in order to keep the increase in global temperature below the benchmark of 1.5 °C. Many methodologies have been proposed, including process change, design change, and efficiency improvement. However, the process sector and fossil fuels account for a sizable share of CO_2 emissions. These emissions often come from stationary locations and in considerable amounts.

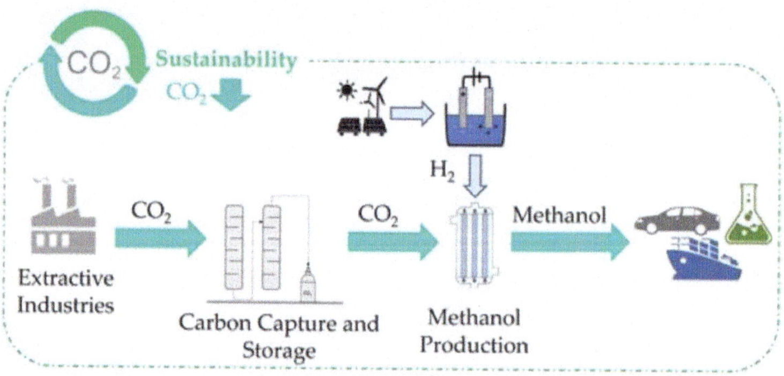

Figure 4.2. Carbon capture.

Carbon capture, (Figure 4.2) use, and storage or CCUS is a potential technology approach to reduce such emissions. It combines three different technological concepts that cooperate to cut carbon dioxide emissions from significant point sources. Technologies that extract carbon dioxide from other gases before or after combustion, such as natural gas, are included in carbon capture. A group of recently developed technologies known as carbon utilization transform the CO_2 that has been captured into useful chemicals, products with additional value, and/or processes raw materials for immediate use. There are many techniques for storing carbon dioxide, including aquifers, impervious rock layers, depleted reservoirs, and underground coal seams. The carbon use and carbon storage industries are less technologically advanced than carbon capture. To achieve the sustainable and environmental aims of decarbonization, however, the technology has to undergo significant technological advancements, social acceptability, commercialization, and widespread deployment.

4.2 ABSORPTION

The most widely used method in the sector is absorption, which uses physical and chemical solvents to take carbon dioxide out of a gaseous mixture. Many types of chemical solvents have been studied and have achieved selected commercialization because of a requirement to reduce energy consumption, avoid solvent degradation, and minimize solvent loss. A good illustration of a physical solvent is selexol. The concept differs when carbon is absorbed via gas hydrates. Adsorption is a different perspective CO_2 collection technology that uses significantly less energy and money than traditional absorption(Jiang et al., 2023). For this technology, zeolites and activated carbon have remained the go-to adsorbents. It has been investigated to modify the adsorbents in various ways. In order to remove CO_2 from ultra-dilute gases like air, adsorbents that undergo repeated cycles of sorption and desorption have become increasingly popular during the past ten years. CO_2 mineralization is the process by which the captured CO_2 interacts with metal cations to produce carbonate minerals. The alkali/alkaline mineral olivine, a magnesium-containing crystal found in igneous rocks, is converted to magnesium hydroxide (brucite) by a complex series of reactions. Under ideal conditions, a pound of olivine may absorb a pound of carbon dioxide from the atmosphere. Utilizing metal organic frameworks to trap CO2 is a revolutionary idea. Membranes are a low-cost carbon capture solution that have undergone significant development but limited CO_2 capture application during the past two decades. They provide solvent-free separation, but more work needs to be done on operational and durability difficulties.

A well-known field, cryogenic CO_2 separation, has seen renewed interest as a result of the necessity to generate revenue from natural gas with high CO_2 levels. The high energy expenses, which are primarily needed to condense all or part of the gaseous mixture, limit the practical applicability. New opportunities for this technology are provided by new cheap classes of materials and novel processing ideas in cryogenic distillation. Another carbon capture technology is calcium looping, often known as the regenerative calcium cycle. It comes after carbonate looping.

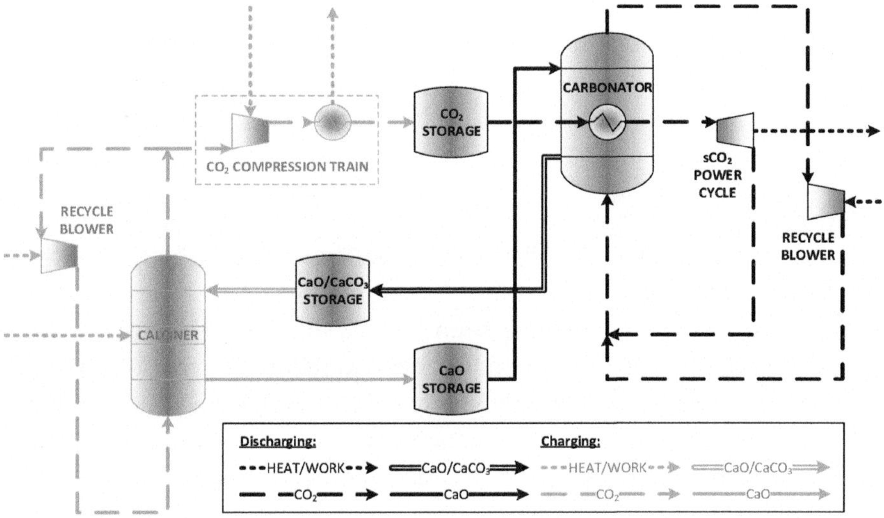

Figure 4.3. Calcium looping.

Carbon dioxide is extracted from a mixture of gases by the direct reaction of calcium oxide (CaO), a sorbent, with carbon dioxide (CO_2) to produce calcium carbonate ($CaCO_3$). To recover the reacting CO_2, the produced carbonate is subsequently calcined in another area of the plant. It is recommended that calcium looping, (Figure 4.3) a dry process with high temperatures (550–1150 °C), be considered as a viable option for the synthesis of cement and hydrogen. In order to make natural gas more marketable, carbon dioxide was initially removed from it in the 1920s, beginning the history of carbon capture. The primary method was absorption using chemical solvents (alkanolamines). One of the reasons absorption is the best developed carbon capture method is because it was the main technology for carbon dioxide separation for the following 50 years(Yang et al., 2008)such as absorption, adsorption, and membrane separation, are thoroughly discussed. New concepts such as chemical-looping combustion and hydrate-based separation are also introduced briefly. Future directions are suggested. Sequestration methods, such as forestation, ocean fertilization and mineral carbonation techniques are also covered. Underground injection and direct ocean dump are not covered.","container-title":"Journal of Environmental Sciences","DOI":"10.1016/S1001-0742(08. The transition of this system to high-pressure systems and testing of other alkanolamines (chemical solvents) during this time period, with the requirement to reduce energy penalty and solvent loss, were the two biggest improvements.

4.3 APPLICATIONS

Due to the commercial application of carbon dioxide in enhanced oil recovery, interest in carbon capture was rekindled in the early 1970s. For instance, the Terrell gas processing plant in Texas (US) was the first commercial carbon capture facility to be built, and it was designed to use captured CO_2 for financial gain rather than as a process difficulty as it had in the past. The facility extracted carbon dioxide from natural gas and sent it to an oilfield in West Texas for better oil recovery. This pattern was copied by a lot of natural gas processing plants in the US. The first chemical processing facility to join the organization was the Koch Nitrogen Enid Fertilizer Plant in Oklahoma (US) in 1982; the captured carbon dioxide was sent to an oilfield in Southern Oklahoma for better oil recovery. ExxonMobil, Chevron, and Anadarko Petroleum collaborated to build a Wyoming carbon capture project in 1986. The carbon capture facility, also known as the Exxon Shute Creek Gas Processing Facility, had a 7 million tonne annual capture capacity(Parker et al, 2011)and from the very beginning, ExxonMobil has successfully implemented several technologies and approaches to effectively manage the substantial volumes of CO2 associated with its production. Many of the technologies and approaches used for managing CO2 at LaBarge are examples of technologies and approaches being proposed for use in carbon capture and storage (CCS (Michael E. Parker, Northrop, Valencia, Foglesong, & Duncan, 2011)and from the very beginning, ExxonMobil has successfully implemented several technologies and approaches to effectively manage the substantial volumes of CO2 associated with its production. Many of the technologies and approaches used for managing CO2 at LaBarge are examples of technologies and approaches being proposed for use in carbon capture and storage (CCS. Of course, the gas was trapped and afterwards used for small home purposes. It was also employed for better oil recovery.

Interestingly, despite numerous advancements in technological and policy development in the Carbon Capture Utilization and Storage (CCUS) industry, the facility is still the largest commercial carbon capture facility globally. In the United States, there were 114 active carbon capture units for increased oil recovery as of 2010. Massachusetts Institute of Technology (MIT) conducted the first comprehensive scholarly study on carbon capture for decarbonization. Beginning in 1989, the Carbon Capture and Sequestration Technologies Program concentrated on the collection, utilization, and storage of carbon dioxide from sources.

4.4 CO_2 EMISSIONS

Carbon dioxide emissions are mostly linked to the burning of fossil fuels and industrial activities like steam-methane reforming and lime-cement production that result in carbon dioxide through chemical reactions. However, burning is where the majority of carbon dioxide emissions come from. Even 40% of the emissions from Small modular reactors (SMRs) and the lime industry come from the direct fuel burning necessary for endothermic processes to take place(McMillan et al., 2016)process-level analysis of industrial heat requirements. The intent is to provide a basis for projecting opportunities for clean energy use. This provides a prospectus for small modular nuclear reactors (including nuclear-renewable hybrid energy systems. With a few exceptions, carbon dioxide capture systems can therefore be generally categorized according to how the CO_2 is separated from the rest of the mixture. Pre- or post-combustion are both possible. As the name suggests, the CO_2 is either separated before the fuel is burned or is caught from the flue gas after the fuel has been burned. Pre-combustion, for instance, is used actively in the natural gas and is also effective in the upgrading of biogas, the purification of hydrogen, and the cleaning of syngas. Post-combustion is typically connected to the treatment of power plant flue gases, although the technology is also useful for industrial CO_2 collection in the paper, cement, iron/steel, and pharmaceutical industries. The kind of process, carbon dioxide concentration, CO_2 purity, and level of necessary carbon capture all influence the technology choice. All carbon capture systems need substantial equipment sizes since the CO_2 concentration in the flue gas from coal-fueled and natural gas–fueled plants are similarly low to the overall volume of flue gas. The low flue gas pressure, which limits the use of high pressure devices like adsorption and membranes, which are more energy efficient than chemical absorption, is another problem with post-combustion. This presents a challenge for the process design because energy-intensive flue gas compression could negate the advantages of energy-efficient adsorption. These limitations are said to be overcome by advancements in adsorbents, but commercial systems are still few. Pre-combustion carbon capture, on the other hand, benefits from intrinsic flexibility because feed gas conditions can vary. Therefore, any technological choice can be used. The amount of carbon dioxide in the fuel, the level of filtration needed, and the gas pressure all influence the technology chosen. For natural gas that has less than 15 percent carbon, absorption is still the preferred option(Cannell, 2003)the EU15 countries and the UK over the next 50–100 years, divided into what are considered: (i.

4.5 CARBON CAPTURE

Direct air carbon capture is a newly developed technique for directly removing carbon dioxide from the atmosphere. The air is passed through a layer of adsorbents and/or membranes, where the CO_2 in the air bonds to specifically developed materials that have a high CO_2 affinity. Some technologies have reached the pilot scale in the process. Alternatives to pre- and post-combustion include chemical looping and oxyfuel combustion. These procedures remove the nitrogen present in the air, leaving practically only oxygen. While the chemical looping process uses a metal oxide as an oxygen carrier to the combustion chamber, the oxyfuel process uses cryogenics and/or adsorbents to separate oxygen. For the purpose of burning fuel, pure (or almost pure) oxygen is used. Therefore, carbon dioxide and water are the primary byproducts of such combustion. Bypassing all other technology alternatives utilized for carbon capture, the purification of carbon dioxide is not a technical limitation because the water can be easily separated. The technology itself, nevertheless, is hampered by large capital expenses. The use of high-temperature metals for confining the heat of reaction is to blame for the majority of these capital costs. The method was formerly taken into consideration as a technology alternative to traditional carbon capture. However, these processes are losing their scientific and economic significance due to significant advancements in alternative technological options and a lack of study in these areas.

Carbon dioxide emissions come from a variety of sources. As a result, a carbon capture system cannot be transferred across applications. For instance, it is not economically feasible to capture CO_2 from coal-fired power plants, but the same technology is becoming accessible for purifying natural gas(Wall, Stanger, & Santos, 2011). As a result, it is impossible to determine the cost of carbon capture on the basis of cost per tonne of CO_2, as it is instead influenced by a variety of other factors, including the presence of other harmful pollutants, safety concerns, human factors, and occasionally even water vapor. Therefore, a comprehensive life-cycle assessment is required. A strong commitment to achieving significant greenhouse gas emissions reductions is preferable to doing nothing from a macroeconomic standpoint. However, the global financial system does not choose or reward governments or firms that act in the interests of all parties. In addition, fierce competition encourages businesses and other organizations to maximize profits. Although governments in most nations control market forces, there is no universally accepted international legislation governing ethical or environmentally friendly behavior. To create policies that impose a cost on adding additional CO_2 emissions to the environment, nations must come to an agreement. As an alternative, society can significantly increase the social value of carbon dioxide. The current inaction on CO_2 emissions is a major factor in the rise in atmospheric CO_2 levels. The severity of the climate issue will increase with continued inaction. Small behavioral changes, such as avoiding single-use plastics or making purchases from green/sustainable companies, can have a significant impact on climate change initiatives.

4.6 LCA FOR CCU/CCS

A tool called life cycle assessment (LCA) was created to evaluate the environmental effects of a technology or product throughout the course of its entire life cycle, from the extraction of resources for the creation of the hardware and materials to their usage and decommissioning. An LCA maps all emissions from these processes and quantifies their contribution to all relevant environmental impacts, from local noise issues to regional acidification and photochemical ozone formation to global climate change. The life cycle includes all the processes required for the technology to perform its function. LCA may detect any issue that occurs when a problem moves from one stage of the life cycle to another or when environmental impacts are traded off due to the systems approach and the extensive coverage of environmental consequences. This makes it the go-to tool for evaluating the environmental sustainability of various carbon capture and storage (CCS) or carbon capture usage methods (CCU)(Full et al., 2023) carbon dioxide removal technologies are urgently needed. The production of biomethane or biohydrogen with carbon dioxide capture and storage are two promising BECCS approaches to achieve these goals. In this study, we compare the advantages and disadvantages of these two approaches regarding their technical, economic, and environmental performance. Our analysis shows that while both approaches have the potential to reduce GHG emissions and increase energy security, the hydrogen-production approach has several advantages, including up to five times higher carbon dioxide removal potential. However, the hydrogen bioenergy with carbon capture and storage (HyBECCS. The economy is a crucial factor to take into account, however, if adopting a particular new CO_2 capture technology causes an increase in overall emissions, the activity is completely pointless if the goal is to assist a net removal of CO_2 from the atmosphere. LCA determines if the system's overall life cycle results in a positive or negative net CO_2 account.

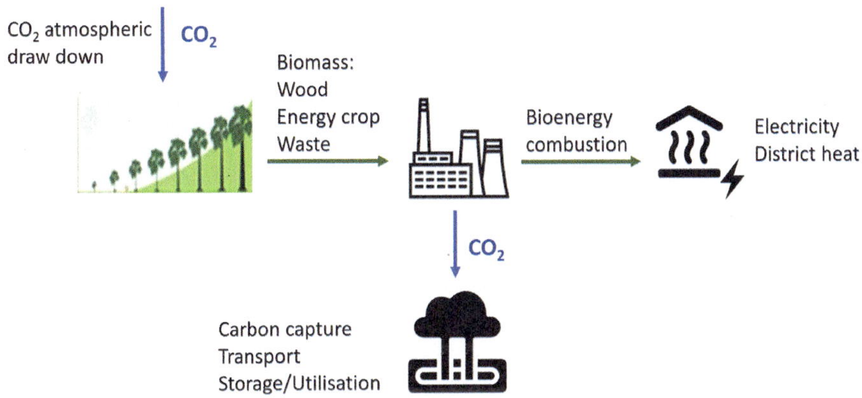

Figure 4.4. Bioenergy with carbon capture.

Bioenergy with carbon capture and storage (BECCS)(Figure 4.4) is one situation that is frequently brought up as an example. A form of BECCS should not be used from an LCA perspective if the biomass used in it was not generated sustainably. Trees shouldn't be utilized in BECCS if they are cut down and replanted, but it takes them 25 years to begin storing CO_2 once more. Such BEECS wouldn't lower the atmospheric CO_2 concentration. In relation to a potential CO_2 capture technique, the same is true. The use of a medium for CO_2 capture should not be taken into consideration if doing so results in increased CO_2 emissions during production, usc, or disposal.

LCA, with its comprehensive examination of environmental effects, also reveals whether a CCS or CCU technique that provides a net benefit for climate change does so at the expense of other detrimental environmental effects, such as the release of harmful chemicals or the use of land in a way that results in a loss of biodiversity. The LCA's perspective is required. The entire life cycle, from manufacture to use and eventual decommissioning, should be taken into account when comparing technologies in terms of their emissions and environmental effects. The World Meteorological Organization (WMO) and the UN Environment Programme (UNEP) both backed a proposal from Malta to create the IPCC during the UN General Assembly in December 1988(Haas, 2000). The IPCC assesses our understanding of climate change. It indicates where there is consensus among scientists on issues relating to climate change and where further research is required. The IPCC ensures objectivity and transparency by producing reports that are produced and reviewed in stages. The IPCC doesn't carry out its own analysis.

4.7 REPORTS

Reports from the IPCC are objective and useful for policy. Five assessment cycles and five assessment reports have been produced by the IPCC since 1988; a sixth report available in 2022. A variety of methodology reports, special reports, and technical papers have also been issued by it. These reports frequently determine the topics of public discussion and have served as a model for many countries implementing climate legislation. The UN Framework Convention on Climate Change (UNFCCC) is in charge of, among other things, intergovernmental climate change negotiations and maintains a database of greenhouse gas emissions from nations throughout the world. In order to meet the climate target established by the Paris Agreement, the UNFCCC is required to keep up with state progress. The Conference of the Parties (COP) meetings, which are heavily covered by the media and attended by key parties, are organized by UNFCCC.

Two exceptional reports stand out among the numerous IPCC reports. The first on the subject of CCS was released in 2005 and describes the advantages and disadvantages of the entire CCS chain in terms of emitter, capture technology, transportation, storage, and CO_2 utilization. In 2005, fossil fuel CCS was the main topic, and the necessity of reducing emissions and recycling carbon received very little attention. The COP conferences in 2017 and 2018 encouraged policymakers to concentrate on limiting global temperature increases to no more than 2°C, with a goal of 1.5°C. As of right now, the temperature increase is around 1°C (2021), which suggests that policymakers have come to terms with the fact that the environment will get much worse in the ensuing decades. The end of 2018 saw the release of a second IPCC report that amplifies the necessity of carbon dioxide removal (CDR)(Keller et al., 2018)threatening severe, pervasive and irreversible impacts. Slow progress in emissions reduction to mitigate climate change is resulting in increased attention to what is called geoengineering, climate engineering, or climate intervention – deliberate interventions to counter climate change that seek to either modify the Earth's radiation budget or remove greenhouse gases such as CO2 from the atmosphere. When focused on CO2, the latter of these categories is called carbon dioxide removal (CDR. There has been a discernible shift from a focus on fossil fuels to the need to achieve a zero-emission strategy by 2050 and a need for negative emissions after 2050 throughout the course of the 13 years between 2005 and 2018. In terms of CCS from biomass combustion, direct air capture and BECCS are the primary methods available for developing a sound climate strategy in the twenty-first century. In its second research, the IPCC predicts that, depending on how quickly society is changing politically, one of four scenarios (P1 to P4) will probably come to pass. P4 requires a startling 22 gigatons of CCS annually for a number of years, whereas P1 has essentially no potential for CCS. We can already tell that P1 won't likely be achieved because major global players don't have a climate strategy.

It is unclear how much CCS will be used, but it is obvious that the amount might range from 1 to 5 GT to 22 GT or more. CCS will be the final resort because it is a tool that cannot be used to save the climate on its own. CCS has been portrayed as a technology that promotes the continued use of fossil fuels throughout the 2000s. This may have been true before the ratification of the decarbonization protocols, but not in a world that is gradually but significantly reducing its reliance on fossil fuels. It is widely believed that once fossil fuels are replaced by other energy sources, carbon capture will become obsolete. But in the near future, this won't be the case. However, to do this, large economies that currently rely heavily on coal must either switch to other energy sources or use CCS. Post-combustion CO_2 capture will significantly lower energy efficiency, even in the most advanced coal-based power plants with energy efficiencies above 45 percent. A 150 MWe coal-based power plant, for instance, has an efficiency of 40.3 percent without CO_2 capture and 29.9 percent with 83.6 percent CO_2 capture, according to a recent NETL research. A report on the cost of oxy-combustion, post-combustion, and IGCC with pre-combustion capture was released by IEAGHG in 2014. In all situations, energy efficiencies were about 35%, with a decrease in efficiency of around 9% relative to no CO_2 capture. A report on the current state of CCUS in electricity production was just released by the IEA.

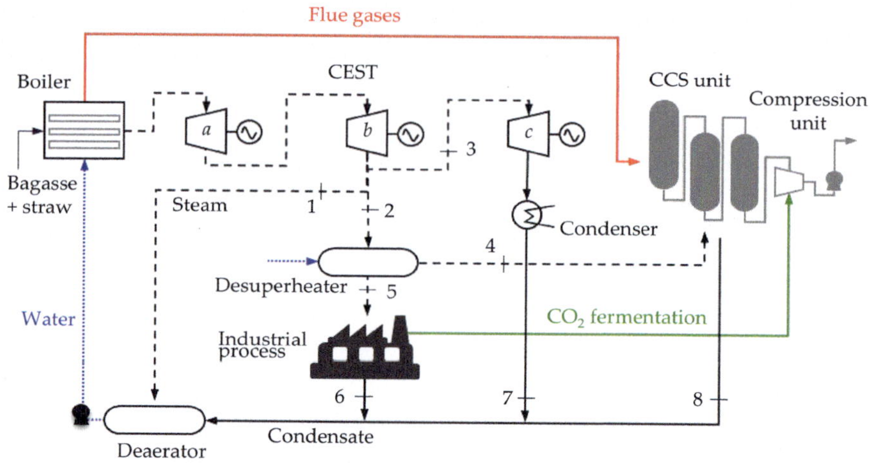

Figure 4.5. Bioenergy with carbon capture-an example of a sugarcane mill in Brazil.

A PCCC pilot plant at the Drax power station, which uses only biomass as fuel, has been set up and is capable of capturing roughly 1.5 tonnes of CO_2 per day. This technique does remove carbon from the atmosphere under some conditions, but if biomass for energy is grown instead of food crops, this could have highly severe consequences. Using agricultural wastes could seem like a better alternative, but it could be in competition with using the waste to enhance the soil and store carbon (an example has been shown in Figure 4.5). The only places where carbon is naturally stored in the earth are the boreal regions. Organic material there oxidizes at such a slow rate that it assembles, yielding moss and ultimately peat. It is crucial to avoid extracting additional peat from the Northern Hemisphere for uses that could oxidize the carbon in it into carbon dioxide (CO_2). The thawing of the Russian tundra, which will release CO_2 and trapped methane, is a terrifying situation. Over the past several decades, direct air capture has been researched. We know from thermodynamics that the low concentration of CO_2 in air makes it energetically more expensive to remove it than exhaust gases. According to estimates, it would take 10% of the world's total energy usage to remove 1 GT annually. This energy must also be clean, producing no new CO_2 emissions. In recent decades, numerous businesses have investigated this possibility, and absorption with non-volatile absorbents is now utilized to keep CO_2 levels in submarines below roughly 1000 ppmv. This might work if low-cost solar heat could be used for regeneration. Unfortunately, carbon capture from mobile sources is not a modest source of emissions, since the transportation industry is a significant contributor to CO_2 emissions. The current trend is to transform fossil fuels into other energy carriers like hydrogen or electricity, with or without carbon capture. Battery-powered vehicles for personal mobility have undergone a revolution, and in Sweden work is underway to install electrical tracks in the road for larger vehicles. In Norway, all new automobiles must be electrically powered by 2025. Even airplanes are powered by electricity, while ferries also use hydrogen. Electricity production has already been suggested, but hydrogen production may be even more intriguing when seen from the perspective of carbon capture.

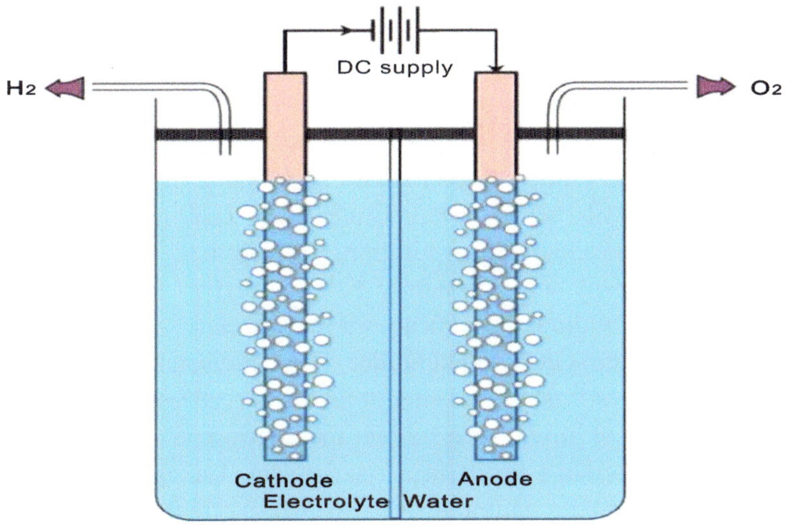

Figure 4.6. Electrolysis of water.

Production of hydrogen might be a wise choice while prices are low. Hydrogen is formed from fossil fuels via steam reforming and shifting to produce CO_2, hydrogen, and trace amounts of other gases. The synthesis gas is produced at high pressure, and CO_2 may be economically recovered by membranes and absorption (blue hydrogen), making them both preferable options. If hydrogen production occurs on a very large scale, the potential to create liquid fuels for transportation is an intriguing element. Hydrogen is produced through electrolysis in a cyclical process, and CO_2 must also be absorbed on a similar scale. This will make carbon capture potentially necessary forever. For more than a century, the necessity for the consumption of fossil fuels has been essential to the world and society. Over the past 50 years, there has been a significant increase in the need for new goods, heating, and transportation. To completely transition away from fossil fuels and toward a sustainable civilization, there is still a long way to go. Particularly in China and India, where achieving a society with no emissions will be extremely difficult over the next few decades, we observe a growing need for fossil fuels. Utilizing CCS to remove CO_2 will be a part of the transition. But society needs to take care. CCS should not be used to increase our reliance on fossil fuels. The only options are sustainable electrification and circular resource reuse, but these approaches will be expensive, and developing nations may need assistance from more developed nations to afford them.

The developed world must concede that most people cannot simply pay 2–10 times more for the same things. The cost of cement for CO_2-free CCS cement will probably double. Electrofuels will cost 3–10 times more under the existing electricity price structure. In any case, without a global CO_2 tax or other equivalent support mechanism for green technologies, it will be exceedingly impossible for any sustainable product to compete with the use of fossil fuels. The corporate social responsibility (CSR) of many fossil fuel-producing corporations contributes to a sustainable climate. One company will frequently be followed by another in a rush to alter names, providing the appearance that society is shifting toward acceptable production. Global stakeholders must understand that when CCS is implemented, the infrastructure cannot just be forced into society without a prior explanation of how it won't affect the environment, or human well-being, and could perhaps result in a risk of CO_2 leakage. Utilizing new technology can stimulate natural apprehensions, particularly if it is highly noticeable in the environment. Pipelines must be installed with the proper materials, knowledge, and safety precautions so that they can demonstrate their safety to those who live nearby. Installation must be done in a way that acknowledges and understands that society doesn't always operate logically. The similar issue pertains to CO_2 storage. For CO_2 storage initiatives to be accepted, it is imperative that CCS-informed states and the research community adequately inform and monitor them.

The CO_2 society has always placed a strong emphasis on technological advancement. The knowledge of CCS in terms of social readiness level (SRL), from stakeholder identification to complete societal adaptation, is also necessary. CCS is now a required technology in many regions of the world in order to meet a 2050 emissions target. For the purpose of storing CO_2, businesses are presently developing pilot projects, full-scale factories, and larger ventures. Norway, the United Kingdom, and the Netherlands are particularly active in Europe. Iceland and Denmark have recently demonstrated efforts for storage locations as well. Many other nations will eventually follow, including France, Spain, Germany, and maybe Italy. Because the true need for change won't arise until the years 2030 to 2050, the rest of the world needs to act but is likely hesitant. The underground CO_2 storage systems must be properly developed as soon as possible in China, the US, and Australia. Due to a simple lack of suitable subterranean storage for CO_2, certain nations are in a scenario where storage is not an easy solution. The transport of CO_2 across borders and across states will be necessary. There are numerous issues with this. To answer these essential issues and ensure the foundation for any infrastructure is in place, significant political change must occur quickly. If the resource is required, the state usually steps in to support the endeavor, assume a sizable risk on investment, and ensure the action is carried out. Similar demands will probably be made of a CO_2 infrastructure. States will initially need to back any attempt to either capture, store, or transform CO_2. Infrastructure becomes crucial at this point. The management of the infrastructure will likely fall to the states as well. To move CO_2 from one hub to the next in a flexible manner, CO_2 infrastructures should be adaptable. Most certainly, the world needs a CO_2 stock exchange that is comparable to and more ambitious than the European emissions trading system (ETS)(Rickels et al, 2021)the emissions cap in the European Union Emissions Trading System (EU ETS. This will enable CO_2 transmission and global trading. Similar to the electricity grid, the CO_2 infrastructure must be able to handle a flexible configuration of both very small and very large consumers. Although it is unknown how this will be handled, it must be considered.

The limitations on battery capacity are the primary distinction between an electrical infrastructure and a CO_2 infrastructure. From the time that CO_2 is caught until it is cleaned, transported, stored, or converted, this refers to the quality and purity of the gas. While some contaminants are largely inconsequential, very few can seriously damage the entire infrastructure. The amount of water is one illustration. Any extra water can damage the pipeline and storage containers, necessitating costly repairs. Sulfur is yet another crucial factor. Any sulfur could potentially harm the catalysis that is most likely already present at a utilization facility (CCU), if the CO2 is intended for conversion there. CO_2 boundary conditions will evolve into a competitive process parameter that also establishes the cost boundary. Whether the CO_2 needs to be purified to ppb levels of an impurity or if it can truly tolerate several percent of particular inert impurities is not irrelevant.

Because of this, a separate infrastructure for low- and high-quality CO_2 may someday be required. Where CO_2 storage facility hubs are located far offshore, the economic factors are fairly significant. It is obvious that this process results in a price increase of several orders of magnitude. The price can drop dramatically, making CCS more appealing and sustainable. If the CO_2 storage can be moved closer to the beach or possibly eventually onshore, the price can also drop significantly. It is most likely irrelevant to the climate whether CO_2 transit makes a sizable contribution to the total emission.

Most likely, a CO_2 hub will also be connected to reuse or energy. Significant energy streams that must be delivered sustainably may be required or formed throughout the conditioning and conversion. It will be difficult to decide how to group CO_2 hotspots into clusters where businesses and states can agree on a set of boundary constraints. Large CO_2 infrastructures that are required in the future may be hampered by these political and technical issues, which might be difficult to resolve(Ball & Wietschel, 2009). The most developed CO_2 separation technology is the reactive chemical absorption, which absorbs carbon dioxide (CO_2) and releases it. Since the 1930s, aqueous alkanolamines have been the standard absorbent for this application. Initial uses of this technology were in a sector where it was necessary to remove CO_2 from a process stream. While the chemical driving force permits removal even as the physical driving force is exhausted, reactive chemical absorption is particularly well suited to situations where little residual CO_2 in the treated gas stream is required. The use of reactive chemical absorbents to reduce CO_2 emissions to the environment from big point sources, such as coal- and gas-fueled electricity generation and heavy industry, such as cement manufacture and steel smelting, has refocused attention on their development in recent decades. In order to reduce past emissions, this approach also involves direct air capture, which presents the added problem of being an extremely diluted supply of CO_2 that is well suited to reactive chemical absorbents. This method offers a way to divert CO_2 emissions from the atmosphere to geological storage when paired with geo-sequestration. In order to attain the 1.5°C global warming targets established as part of the Paris Agreement international treaty, carbon capture and storage has been acknowledged as an essential technology.

Additionally, it offers a source of CO_2 for applications including the production of liquid fuels and building materials. The drive to better understand the chemical and physical properties of reactive chemical absorbents and the need to reduce capital and operating costs, increase absorbent longevity, and lower the risk of emissions are all driven by the use of reactive chemical absorbents to massively reduce CO_2 emissions to the atmosphere. We must appreciate the molecule, its geometry, and especially its electrical structure in order to fully understand the chemistry of CO2. As seen in Figure 2.1, CO2 is a linear, highly symmetric molecule. Strong double bonds connect the two oxygen atoms on either side of the core carbon atom. Positively charged reactants tend to interact with the partial negative charge on the oxygen atoms, while negatively charged reactants tend to attack the positive charge of the central carbon atom. The next step is to separate the two distinct types of processes that can affect the CO_2 molecule. The first kind is referred to as a redox reaction. In a redox reaction, the oxidation state of thc carbon atom is reduced from +4 to a lower oxidation state, such as +3 in formic acid (HOCO), +2 in carbon (CO), 0 in carbon (C), 2 in methanol, or even 4, which is the oxidation state of the carbon in methane (CH_4). While none of these redox processes are necessary for CO_2 capture applications, they are crucial for CO_2 consumption. They are typically slow and difficult to stop.

4.8 SUBSTITUTION REACTIONS

Substitution reactions, which leave the formal oxidation state of C at +4, are easily reversible processes important for CO_2 absorption via chemical reactions(Garrido-Barros et al, 2017). To be effective, the reactants must have a partial or complete negative charge that will react with the positive charge of the carbon in the center of each of these reactions, which all entail substitution at the central carbon atom. Nucleophiles and/or Lewis bases are molecules with a partial negative charge that are drawn to a positive charge. Electrophiles and/or Lewis acids are the positively charged (full or partial) reaction partners that would react with the partially negatively charged oxygen atoms. Of course, the proton, H_+, is both a Lewis acid and a Bronsted acid. R1 and R2 are carbon-based chains in secondary amines. In protic solvents like water, the proton lost from the amino group will typically be associated with a second amine molecule or result in the formation of carbamic acid in aprotic solvents where protons do not readily exchange between molecules. The relative basicity of the carbamate, amine, and other Bronsted bases present, as well as the proton capacity to migrate through the solution, determine where the proton finally resides. Because there is no proton available to act as a leaving group and the reaction of replacing an N-C bond with another N-C bond would be extremely slow, tertiary amines cannot directly react with CO_2. However, they are still bases and can take in protons that are released during reactions involving CO_2 and other molecules. A number of other aspects are crucial for an amine molecule to perform as a successful CO_2 absorbent, in addition to its reactivity with CO_2 and the yielded absorption capability.

The absorbent is frequently exposed to additional components such as O_2, SO_x, NO_x, H_2S, and metal-containing particles in the gas environments from which CO_2 is absorbed. The absorbent is also subjected to high temperatures in methods for thermal based CO_2 capture. In these operational situations, long-term stability is crucial. Thousands of tonnes of amines will be used in any large-scale deployment. Avoiding the use of substances that are highly toxic to humans, animals, or the environment is preferable, as is the use of substances that biodegrade rather than bioaccumulate, given that some exposure of workers to the amines is likely to occur, and some emissions to the environment will undoubtedly occur. Aqueous solutions of one or more amines, including but not limited to alkanolamines, are the most popular amine-based solution for CO_2 absorption(Arcis et al, 2014).

At low temperatures, it has a higher CO_2 absorption capacity, while at high temperatures, it exhibits a similar capacity to MEA, leading to a higher cycling capacity. It has been identified as a single amine that is combined with 2-(2-aminoethylamino)ethanol in this formulation. The fact that it is a secondary amine may make it more likely to create nitrosamines when NOx, which may be present in some flue gases, is present. Additionally, little is known regarding its long-term stability and degradation. 2-(Ethylamino) ethanol (EAE) is an amine that structurally resembles MAE and has characteristics in common with it. The primary distinction is that it has been employed as an absorbent rate enhancer in formulations with the tertiary amines 3-(dimethylamino)propan-1-ol and diglycolamine. It is also known to react with CO_2 very quickly. Its secondary amine status and potential for nitrosamine production are drawbacks, and nothing is known regarding its stability or degradation. The steric hindrance increases its capacity to absorb CO_2 and its capacity to cycle, but it slows down its rate of CO_2 reaction in comparison to EAE. It has been successfully used in small-scale pilot plants, and an evaluation of the thermal stability indicated that it was marginally superior to MEA. It is likewise a secondary amine, although it has not been researched if steric hindrance reduces its propensity to create nitrosamines. Like MEA, 2-(methylamino)ethanol (MAE), 2-(ethylamino) ethanol (EAE) and 2-(isopropylamino)ethanol (IPAE) have low viscosities but are completely miscible with water and don't precipitate. Diglycolamine

(DGA) has been employed for CO_2 capture applications for a while, and interest in the compound has been constant from 2010 to 2020(Pant, 2020). DGA has previously been used as a single aqueous amine for low pressure but high CO_2 fraction applications in natural gas sweetening. The more recent articles concentrate on its use in biogas upgrading applications, which have a high CO_2 component composition but a similar low pressure. Diisopropanolamine (DIPA), which is extensively described and utilized in process modeling software, has also been used for high-pressure CO_2 capture applications. Recent DIPA research has concentrated on defining its formulation with different amines, usually to increase slow mass transfer performance while keeping high-pressure applications in mind.

The amine 2-amino-2-hydroxymethyl-1,3-propanediol is sterically hindered and has a structure similar to AMP, but each methyl group linked to the alpha carbon has an alcohol group added to it. Its prominence is peculiar because the additional alcohol groups limit the capacity at low CO_2 partial pressure and make it more suitable for high-pressure CO_2 removal applications by reducing the basicity of the amino group due to their electron-withdrawing effect. Surprisingly, the extra alcohol groups do not seem to have a particularly negative effect on viscosity, and the increased hydrogen bonding ability in an aqueous solution results in a low vapor pressure. To attain practical mass transfer rates in the majority of applications, formulation with a faster-reacting primary or secondary amine that is not sterically inhibited will be required. Although little is known about its stability, its thermodynamics are typical of sterically hindered amines that do not generate carbamates. Tertiary amines are most commonly found in mixtures with primary or secondary amines to attain mass transfer rates. Despite having very comparable structural characteristics, Diethylethanolamine (DEEA) is more fundamental and favors faster CO_2 mass transfer. Its main selling point is that it is more basic than both amino-2-methyl-1-propanol (AMP) and methyldiethanolamine (MDEA), which improves mass transfer and absorption capacity. It also has a lower CO_2 absorption enthalpy and viscosity. However, in comparison to AMP, it has a lower thermally driven CO_2 cycle capacity and is less stable. As a somewhat reliable and inexpensive amine for formulation with primary and secondary amines, DEEA offers an alternative to AMP, however, its different thermodynamic properties would dictate different operating conditions.

1-Dimethylamino-2-propanol, often known as DMA2P or 1DMAP, is another tertiary amine that has recently attracted a lot of research attention(Liang et al., 2015)the CO2 absorption performance of aqueous

1-dimethylamino-2-propanol (1DMA2P. Its structural similarity to DMEA is due to the methyl group at the 2-position in place of the hydrogen. Although it is close to the alcohol group in the structure, this methyl group is unlikely to interact with the amino group. It shares similarities in basicity with the previously mentioned structurally related tertiary amines. Similar in structure to 1DMAP, DEAB (4-diethylamino-2-butanol) has ethyl substituents on the amino group and an extra carbon in the chain separating the amino and alcohol functionalities. Greater basicity compared to DEAB and the other amines already mentioned is the overall result of the closer proximity to the electron-withdrawing alcohol (pKa of conjugate acid >10.2 at 25°C, compared to 9.55 for 1DMAP). As a result, it has a higher capacity for absorption and a quicker mass transfer behavior than other tertiary amines. To achieve relevant mass transfer, however, other primary or secondary amines must still be used in formulation. About its thermodynamics, and consequently its cyclic capacity or stability, little is now known. Because each polyamine amino functional groups are CO_2-reactive, this is a benefit. The same molecule may theoretically produce several carbamate functional groups, accept multiple protons, or form a carbamate and accept a proton. The characteristics of each amino group and how close they are to one another determine how much of this actually happens in practice. If two amino groups are close to each other, carbamate production or protonation of one will affect the reactivity of the other. The amino groups may also communicate through intramolecular interactions, such as when a carbamate is stabilized or a proton transfer occurs more quickly due to intramolecular ring formation. The fact that acyclic aliphatic polyamines have many reactive amino groups also makes them more susceptible to oxidative and thermal breakdown.

4.9 ETHYLENEDIAMINE (EDA)

In terms of applications, ethylenediamine (EDA) is by far the most developed of the acyclic polyamines. It is structurally MEA with a second amino group in place of the alcohol group. It has been employed in pilot-plant testing and is still the subject of basic research into the properties of linear diamines. Although EDA is more basic and reacts with CO_2 more quickly than MEA, the amino groups exhibit characteristics that are similar to those of MEA. Since the amino groups are close together, protonation of the second amino group happens at a pH level below what is necessary for CO_2 uptake after the first amino group has been protonated. It can, however, result in a dicarbamate. The usefulness of EDA is constrained because it is less stable and more volatile than MEA. Hexamethylenediamine (HDMA) usage has generated interest. While there are six rather than two carbons in the chain that separates the amino groups, it is comparable to EDA. With tertiary or sterically hindered amines, HDMA may operate as a rate promoter since it is somewhat more basic than EDA, has a higher absorption capacity, and reacts with CO_2 more quickly. To ascertain its stability and cycle capability, more research is required.

Many studies have focused on the amine 3-(Methylamino)propylamine (MAPA). It is comparable in basicity to EDA and has one more carbon atom between the amino groups. A methylation of one of the groups results in MAPA having both a main and a secondary amino group. Although the second protonation constant is substantially smaller than the first, it still falls within a reasonable range for CO_2 absorption. It is more fundamental than EDA. MAPA has demonstrated quick CO_2 reaction kinetics and strong cyclic capacity. In formulations with other tertiary and sterically complex substances, it has been generally regarded as suitable.

Because of their high viscosity, sarcosine and captured-hindered amines cannot be used alone at high concentrations. It has a strong cycle capacity and well-characterized thermodynamics. Its stability appears to be a drawback because it breaks down quickly when oxygen is present. Due to the potential for nitrosamine production, the inclusion of a secondary amino group further restricts its utility in the presence of NOx. 2-(2-Aminoethylamino)ethanol, often known as AEEA, has traits in common with both EDA and MAPA. It has a main and secondary amino group like MAPA and has the same distance between amino groups as EDA. The substitution of ethyl alcohol is what makes a difference. The closeness of the amino groups makes it similar to EDA in that only the first protonation constant is important for CO_2 uptake. The second happens when the pH is too low. Additionally, the inclusion of the alcohol group reduces the basicity in comparison to MAPA and EDA. AEEA reacts with CO_2 quickly, but not as quickly as HDMA, probably because of its lower basicity. The majority of investigations have concentrated on CO_2 reaction kinetics and vapor-liquid equilibria. Regarding its stability, thermodynamic, and cyclic capacity characteristics over a wide temperature range, little is known. The only triamine found among the linear polyamines is bis-(3-aminopropyl) amine (APA). It has received less research attention than the other polyamines already covered. A core secondary amine and two propylamine substituents are attached to its symmetrical structure. The separation of the amino groups means they all have the potential to be active proton acceptors and carbamate formers. It also has low viscosity and reacts quickly with CO_2. To completely assess the performance of the material as an absorbent, further information must be gathered, particularly as a function of temperature.

4.10 CYCLIC AMINES

This definition of "cyclic amine" refers to an amine compound that has a cycle. The cycle might or might not include the amine itself. These comprise mono- and diamines, and all but one of them are aliphatic; the only exception is an aromatic monoamine. The fact that cyclic structures need less energy than their acyclic counterparts is the main factor driving the adoption of cyclic structures in molecules. Because opening the rings of cyclic compounds is more difficult than cleaving or substituting acyclic molecules, it makes sense that cyclic molecules would be less prone to deterioration. PZ has been extensively examined over the past ten years and has a stable and quick reactivity with CO_2(Vaidya & Kenig, 2007). Although it has two secondary amino groups that, in the presence of NOx, can produce nitrosamines, its stability permits CO_2 stripping at higher temperatures, which also eliminates any nitrosamines that may have formed. In order to comprehend the characteristics of PZ better, 1-methyl-piperazine (1-MPZ) has frequently been used. It preserves the quick reaction with CO_2 and possesses secondary and tertiary amino functionality. It is less basic than PZ. Although less stable than PZ, it is still among the more stable amines that have been researched thus far. It does appear to have a lower capacity to absorb CO_2, which is most likely caused by its inability to produce a dicarbamate. The compound, 2-methyl piperazine (2-MPZ), is extremely similar but preserves two secondary amino groups, one of which is sterically hindered due to the placement of the methyl group.

CHAPTER 5

Biodiversity and Carbon Capture

CONTENTS

5.1 INTRODUCTION

Many of the globalbiogeochemical cycles have undergone profound changes in recent centuries as a result of human activity. The altering of the world's carbon cycle (Figure 5.1) was one of the earliest and most obvious of these changes.

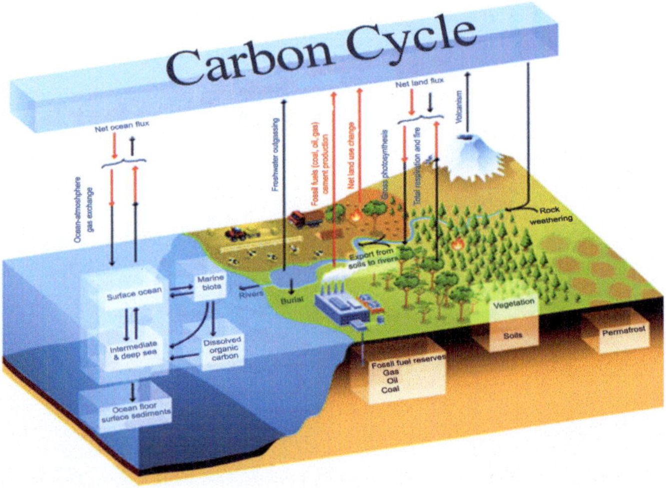

Figure 5.1. Carbon cycle.

The amount of carbon dioxide (CO_2) has increased by 31% from a preindustrial concentration of about 280 ppm to 368 ppm in 2000 due to the significant release of carbon stored by both vegetation and prehistoric ecosystems (i.e., fossil fuels)(Swinglandet al., 2002)Prance, Saunders, Malhi, et al., 2002 (Swingland, Bettelheim, Grace, Prance, Saunders, Malhi, et al., 2002)Prance, Saunders, Malhi, et al., 2002. In the last 420,000 years and probably not in the last 20 million years has this concentration been exceeded. Additionally, the rate of growth in CO_2 concentration during the past century is at least a factor of ten higher than what the planet has experienced over the previous 20 years. The most significant greenhouse gas that is increasingly absorbing solar heat and warming the planet is CO_2. This additional CO_2 may affect terrestrial ecosystems in a variety of ways, including warming the climate, accelerating plant growth and biomass, and changing the species' competitive relationships to change the composition of the ecosystem. In their pre-agricultural state, forests were thought to have covered about 57 million km^2, contained about 500 PgC in living biomass, and contained a further 700 PgC in soil organic matter, holding more than twice as much carbon as the preindustrial atmosphere overall and recirculating about 10% of atmospheric CO_2 into the biosphere each year through gross photosynthesis. Since the beginning of the agricultural revolution 8000 years ago, there has been a global decline in the amount of forest cover; nonetheless, by the year 1700, only around 7.5% of the world's forest area had been lost. Since the industrial revolution, the degree and scope of human alteration of the biosphere have increased, and by 1990, 20 to 30 percent of the original forest area had been lost. The reduction in forest cover is responsible for 45% of the rise in atmospheric CO_2 since 1850. Although fossil fuel emissions of carbon have recently overtaken those from deforestation, (Figure 5.2) land use change still accounts for around 25% of all current human-induced carbon emissions.

Figure 5.2. Deforestation in the Chiquitania Region, Bolivia.

During this period, known as the Holocene, population and activities significantly grew, but, up until the Industrial Revolution, atmospheric CO_2 concentration slightly fluctuated, rising generally from a post-glacial minimum of 260-285 ppm in the late 19th century. The global carbon cycle is assumed to have remained roughly constant given the relative stability of CO_2 concentrations during the last few millennia. There are four main carbon stores: the geological, oceanic, terrestrial, and atmospheric reservoirs. Changes in the atmospheric carbon store, despite being the smallest component, are ultimately what establish a direct connection between variations in the global carbon cycle and variations in climate. Under normal conditions, these main channels for the transfer of carbon from the atmosphere to land or oceans are roughly in balance over an annual cycle and account for an exchange of 210 PgC per year, of which the land contributes the majority(Schaphoff et al., 2006). The total amount of carbon emitted into the atmosphere every year as a result of human activity is more than 25 times more than this yearly exchange. Climate has an impact on the biological and physical systems that control this exchange. As a result, a slight change in climate (change in temperature) could cause them to shift in a way that is proportional to the amount of emissions linked to human activity that caused them.

5.2 ORGANIC CARBON

The outflow of dissolved organic carbon (DOC) through rivers to the sea balances the transfer by plants of 0.4 PgC/yr of inert carbon from the atmosphere to the soil(Schwab et al., 2022). The weathering of rock carbonates, which react with atmospheric CO_2 to generate dissolved inorganic carbon (DIC), contributes a flux of dissolved inorganic carbon (DIC) of 0.4 PgC per year to this outflow. All of the DOC and half of the DIC are eventually respired to the atmosphere as part of the 0.8 PgC/yr flow through rivers to the sea. This leaves 0.2 PgC/yr to be deposited as fixed carbonates, the remains of extinct marine animals, in deep-sea sediments. Rocks made of carbonate are formed from these deposits. In even longer time periods, carbon dioxide is released into the atmosphere as a result of tectonic action, such as volcanism, and organic matter is buried as fossil organic matter. On geological time spans, these longer time-scale processes have a significant impact on atmospheric CO_2 concentrations, but they have little impact on the time scale associated with human expansion. At least two very early periods, 400 and 200 Myr BP, presumably saw exceptionally high concentrations of more than 3000 ppm. CO_2 was taken out of the atmosphere when photosynthetic equipment developed and took over on a worldwide scale, and geological evidence suggests that concentrations were less than 300 ppm by around 20 million years ago. Therefore, it is likely that atmospheric CO_2 concentrations now are substantially higher than they have ever been in the past 20 million years. We have a comprehensive understanding of the previous four glacial cycles spanning a time span of 420 kyr because of the accurate measurements of the CO_2 record from the Vostok glacier. According to the findings, atmospheric CO_2 concentrations fluctuated between 180 and 300 ppm, with lower amounts during glacial epochs. Because more carbon is probably stored on land during interglacial oceanic processes that release carbon must account for the reported greater atmospheric CO_2 levels during these times. However, coincidental changes in CO_2 concentration have played a key role in locking-in these 100 kyr cycles by amplifying their effects rather than initiating them.

5.3 CARBON STORAGE

The partitioning of carbon storage on land occurs between the soil and the vegetation. Over 75% of all terrestrial carbon stores are found in soils globally, though the percentage varies depending on latitude and land use. With about 47% and 25% of the world's total carbon storage, respectively, forests and woody grasslands/savannas are by far the largest carbon sinks. Other ecosystems typically keep only a small amount of aboveground carbon, with soil stocks ranging from 100 to 225 PgC. Carbon is accumulated in forested ecosystems (Figure 5.3) by absorbing atmospheric CO_2 and assimilating it into biomass.

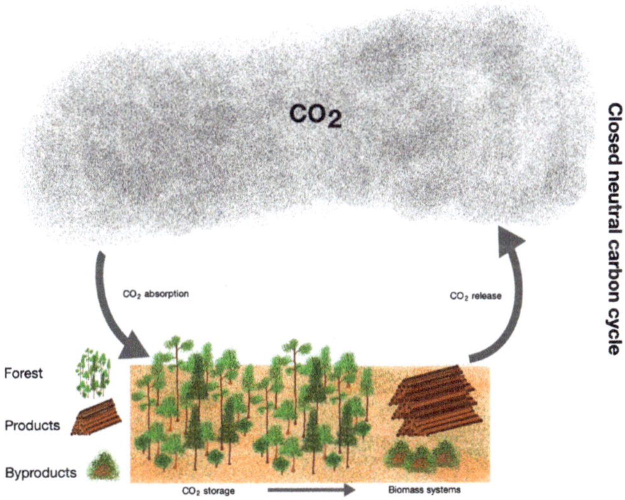

Figure. 5.3: Carbon and the forest (Source: Nunes LJR, Meireles CIR, Pinto Gomes CJ, Almeida Ribeiro NMC. Forest Contribution to Climate Change Mitigation: Management Oriented to Carbon Capture and Storage. *Climate*. 2020; 8(2):21. https://doi.org/10.3390/cli8020021)

In a forest ecosystem, carbon is stored in a number of different pools, including litter, woody debris, soil organic matter, and forest products. Carbon constitutes up to 50% of the dry biomass of trees. Any activity that alters the biomass in soil and vegetation has the ability to either store carbon from the atmosphere or release it into it. Boreal forests are the only terrestrial ecosystems that produce more carbon overall (26%), followed by tropical and temperate forests at 20% and 7%, respectively. However, the locations where carbon accumulates vary significantly between different types of forests. In tropical forests, the total amount of carbon is distributed fairly evenly above and below ground, but up to 90% of the carbon in boreal ecosystems is stored in soil. The main cause of this discrepancy is temperature, which, while encouraging rapid decomposition and subsequent nutrient recycling at low latitudes, inhibits slow decomposition and subsequent nutrient recycling at high latitudes. Additionally, only a small part of the carbon in wetlands is found in plant biomass due to sluggish breakdown rates in peaty soils and other habitats with high carbon concentrations(Aerts, Verhoeven, & Whigham, 1999).

A byproduct of plant respiration, or between 30 and 50 percent of the total amount of carbon taken by vegetation, is released back into the atmosphere. Net primary production refers to the leftover carbon that is fixed as organic matter above or below the surface. Depending on the climate, soil type, and species composition, different vegetation types have different NPPs. Seasonality limits its value in broadleaf temperate forests, and this effect is particularly pronounced at higher latitudes even if these woods are extremely productive for a portion of the year. It is obvious that different vegetation types process carbon at varying rates. By dividing the total stock by the NPP, this value, expressed as the typical residence time for assimilated carbon, can be calculated. The outcome differs for vegetation and soil. With regard to plant biomass, there is no clear correlation between NPP and, within non-crops ranging from 3 to 22 years. However, there is a significant correlation between NPP and for soil: in tropical forests, where NPP is high, carbon residence time in soil is only around 10 years, but residence durations are much longer in colder habitats, where NPP might be an order of magnitude lower. Focusing just on plant biomass reveals that the sink capacity of various vegetation species is related to the sum of their NPP and residence duration, which also happens to be proportional to the biomass carbon stock at the time.

5.4 CARBON CYCLE

The global carbon cycle in a quasi-equilibrium condition persisted for the majority of human history and the entire Holocene. But human economic activity and population have greatly increased in recent centuries. Even while these activities had significant negative consequences on the local ecosystem, such as widespread urban and industrial pollution, until recently it was believed that the Earth as a whole served as a vast source of raw materials and a sink for waste. In the latter half of the 20th century, this viewpoint started to shift. The emergence of a measurement that served as a measurable indicator that humans were changing the Earth's fundamental biogeochemical cycles on a global scale was one of the most significant factors influencing this shift in thinking. The statistics demonstrate that, despite certain seasonal and interannual changes, atmospheric CO_2 concentrations have climbed inevitably. They show that the global carbon cycle, one of the most fundamental of the large global cycles, is experiencing disruptions from its quasi-equilibrium Holocene state. One of the earliest signs that biogeochemical cycles were being disturbed on a global scale was this carbon disruption, and since then, evidence of similar disruption to other biogeochemical cycles has surfaced. For instance, mechanized fishing removes >25% of the primary production of the oceans in some regions and 35% in the temperate zones, with the release of SO_2 to the atmosphere by coal and oil burning being at least twice as large as the sum of all natural emissions. Additionally, more nitrogen is now fixed synthetically and applied as fertilizers in agriculture than is fixed naturally in all terrestrial ecosystems.

A new geological epoch called the Anthropocene, with an assumed beginning date in the late 18th century, coinciding with James Watt's creation of the steam engine in 1784, has been postulated as a result of such widespread upheaval(Locher & Fressoz, 2012). This signifies a fundamental shift in perspective and the realization that modern human activity and the operation of the Earth system are now intertwined. Although an increase in CO_2 may have an immediate impact on terrestrial ecosystems, its potential to be a major greenhouse gas causing global warming is perhaps its most well-known impact. These processes had released 270 Pg of carbon by the year 1998. When fossil fuels are burned, the carbon that was initially stored by the biosphere and later transferred to geological reservoirs effectively being removed from the rapid carbon cycle is released back into the atmosphere. The direct transfer of carbon from the biosphere to the atmosphere, primarily as a result of the conversion of forests into farmland and pasture, has been the other major source of carbon in the contemporary carbon disruption. Although this process predates human civilization, it has recently experienced a sharp acceleration in its rate of change. Most human cultures have relied on altering natural landscapes since the discovery of fire control, which has led to changes in the carbon storage densities of forests, savannahs, and grasslands. Since 7000 years ago, the majority of temperate forests in Europe, China, and India's monsoon and dry forests have been gradually removed to make way for pasture and crops. Only a small portion of the original forest area survived into the industrial period.

To calculate the areas of natural vegetation cover lost to crops and pasture, researchers created a spatially explicit historical database of potential vegetation cover and historical land use. By 1700, steppes, shrublands, and forests had all experienced losses of about 7%. These processes have accelerated significantly since the industrial revolution and the age of European global colonization, and by 1990 forest area had reduced by 30%, steppes/savannas/grasslands by 50%, and shrublands by 75%. The destruction of seemingly unaltered natural habitats, which has also resulted in a significant release of carbon, is not included in these statistics because it is not part of the total conversion to pasture and cropland. There is more uncertainty because deforestation, logging, and regrowth are harder to track than industrial activity and the net carbon emissions from deforestation are harder to measure. Over time, there are remarkably different changes. In contrast to the tropics, where deforestation has increased significantly since the 1950s and is now responsible for almost all current emissions, temperate deforestation rates have significantly reduced and there has even recently been a net expansion of forest cover in North America and Europe. The primary land-use changes that have an impact on carbon storage are: the permanent clearing of forest for pastures and arable crops; shifting cultivation, which may vary in scope and intensity as populations rise or fall; logging with subsequent forest regeneration or replanting; and abandoning agriculture and replacing it with secondary forest regrowth or planting i.e., deforestation, afforestation and reforestation. Although the ultimate result has been a loss of carbon from forests, several of these processes like shifting agriculture, logging, clearing for grazing, and abandonment entail interactions between forest degradation and later recovery.

5.5 A GRAVE ISSUE

The expansion of croplands at the expense of natural ecosystems has predominated and continues to predominate the net efflux of carbon from the terrestrial biosphere, with the tropical forests of Southeast Asia (0.76 PgC/yr in 1990) and Africa (0.34 PgC/yr in 1990) exhibiting the highest levels of activity(Huang & Gerber, 2015). Southeast Asia accounted for the majority of net emissions from wood harvesting in 1990 (0.19 PgC/; 66%

of total emissions) and China (0.07 PgC/yr; 23% of total emissions). Due to indigenous populations in the Americas and Southeast Asia abandoning agricultural regions, shifting agriculture was declining in the 19th century. But since the middle of the 20th century, growing population demands have caused cultivation areas to grow and rotation times to shorten, making shifting agriculture an increasingly important source of carbon. A carbon sink has developed as a result of the increasing abandonment of agricultural lands and the recovery of forests in North America and Europe. The FAO may be overestimating land-cover change, according to recent research in a number of nations. On the other hand, some processes may be increasing carbon emissions but are not taken into account in estimates. By the end of 2000, approximately 480 GtC had been released into the atmosphere as a result of human activity (91 percent of this since 1850, 50 percent of this since only 1968). In the 1970s, the total contribution of burning fossil fuels merely overtook that of changing land use, but today, fossil fuel combustion is responsible for 75% of current emissions(Hansen et al., 2013).

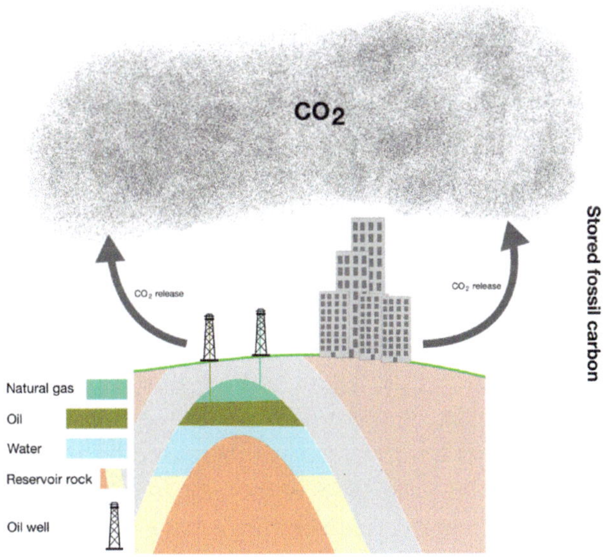

Figure 5.4: GHGs (Source: Nunes LJR, Meireles CIR, Pinto Gomes CJ, Almeida Ribeiro NMC. Forest Contribution to Climate Change Mitigation: Management Oriented to Carbon Capture and Storage. *Climate*. 2020; 8(2):21. https://doi.org/10.3390/cli8020021)

The Northern Hemisphere does have slightly higher concentrations than the Southern Hemisphere, which serves as a planetary-scale smoking gun that serves as a reminder of the high rates of fossil-fuel combustion in northern mid-latitudes. Although net emissions to the atmosphere fluctuate annually primarily due to changes in oceanic and terrestrial uptake that lie beneath the continuous human-related outflow of CO_2, the rate of increase in atmospheric CO_2 is dramatic: it has averaged around 0.4 percent per year since 1980. Because of lower terrestrial uptake in tropical ecosystems experiencing elevated temperatures, droughts, fires, and clouds, the effects of El Nino occurrences, for instance, can briefly result in high rates of CO_2 emission to the atmosphere. Although the biosphere may be directly impacted by an increase in atmospheric CO_2, the major reason to be concerned about rising CO_2 is its function as a greenhouse gas (GHG). GHGs allow some of the long-wave thermal radiation that is sent back out into space to be absorbed, but they prevent some of the short-wave solar radiation from entering the Earth's atmosphere. Despite not being the most efficient GHG, CO_2 is present in quite high concentrations and hence accounts for the majority of the radiative forcing caused by all GHGs. Other GHGs are better at absorbing radiant energy, like methane and nitrous oxide. Calculating CO_2 equivalents, the warming potential of each gas relative to CO_2 over a given time period, typically 100 years allows for comparison with CO_2. Although their atmospheric quantities are substantially lower, methane (CH_4) and nitrous oxide (N_2O) both have warming potentials that are 23 and 296 times more than that of carbon dioxide (CO_2).

5.6 TEMPERATURE PATTERNS

The temperature pattern also displays the relatively constant atmospheric CO_2 content during the millennium leading up to the end of the 19th century. Around 1900, a gradual but persistent cooling that had been evident since 1000 was abruptly broken, and temperatures have risen ever since. Since 1861, the average global temperature has increased by 0.6°C (0.2°C, 95% confidence interval). The year 1998 is thought to have been the warmest on record, and significantly, temperatures have risen high enough to surpass the uncertainty boundaries associated with proxy measurements taken for historical times. The amount of CO_2 in the atmosphere has also sharply increased over the same time period. Sixty percent of all GHG-induced warming is now considered to be caused by the 31 percent rise in atmospheric CO_2 since 1750. Therefore, it is not surprising that the amount of CO_2 that is predicted to be released into the atmosphere has a significant impact on the modeling of climatic scenarios over the next 100 years. Since three-quarters of all CO_2 emissions are now attributed to fossil fuel burning, the socioeconomic model that underpins estimates of fuel consumption has a significant impact on future climatic scenarios. The IPCC (2001) predicted that future temperature rises over the next 100 years would range from 1.5 to 5.8 °C and that the atmospheric CO_2 concentration would climb from its current level of 368 ppm to between 500 and 1000 ppm(Thompson & Malhi, 2002). These modeling results roughly predict that every 3 Pg of CO_2 emissions will result in an additional 1 ppm of atmospheric CO_2 by 2100, and that each additional ppm will result in an increase in world mean temperatures of less than 0.01 °C. These conversion factors are subject to assumptions regarding the temporal pattern of emissions and the future behavior of the ocean and terrestrial carbon cycles, but they offer a basic method for calculating the climatic effects of alternative carbon mitigation scenarios.

Despite being noticeable, the warming since the 19th century does not accurately reflect the overall amount of CO_2 generated as a result of human activity. Due to reabsorption by the oceans and land surface, only a small portion of this amount is ultimately added to the atmospheric CO_2 stock. The division of the net terrestrial sink into tropical and temperate zones, as well as between human activities and natural carbon sinks, is less certain. The division between tropical and temperate zones is based on research on the CO_2 distribution in the atmosphere. The main uncertainty in this budget stems from the estimation of emissions related to land use change; if these are exaggerated, the terrestrial carbon sink will also be overestimated. Currently, industrial activity and tropical deforestation emit about 8.1 PgC per year as CO_2, yet only 40% of this (3.2 PgC per year) contributes to the buildup of CO_2 in the atmosphere. This 40 percent is what adds to the radiative forcing of the Earth's atmosphere, but in order to comprehend how this radiative force will change in the future, it is equally crucial to comprehend what will happen to and how permanent the remaining 60 percent will be. The oceans contain around 50 times more carbon than the atmosphere does. Large-scale exchange between the events over a period of hundreds of years, but because of the strong solubility and chemical reactivity of CO_2 in water, exchange at the sea-air interface can happen quickly and on a considerably shorter time scale. The oceans take in 1.7(0.5) PgC per year from the atmosphere. The precision of this estimate is relatively good despite the uncertainty in a variety of processes, demonstrating agreement between model outputs, values obtained from observations of atmospheric O_2 and C, and scaled-up *in situ* measurements.

5.7 NATURAL MECHANISMS AND CO_2

Natural mechanisms involving both physical and biological processes mediate the movement of CO_2 into or out of the oceans(P. W. Boyd & Ellwood, 2010). The direction of molecular diffusion across the sea-air interface depends on the difference in the partial pressure of CO_2 in each medium. Modeling of this transfer rate into water shows that it varies on sea temperature, salinity, and pH in addition to wind speed. One percent of the CO_2 is left in its original dissolved, non-ionic form after 99% of it reacts chemically with water to form bicarbonate and carbonate ions. The carbon may subsequently undergo additional chemical processes or be taken up by phytoplankton and converted into biomass. Particulate organic carbon produced by photosynthesis in the waters sinks to great depths. The majority of this exported carbon eventually makes its way back to the surface via the underlying marine circulation and is often released as CO_2 far from its source. The impact of this biological pump mechanism is significant: without it, atmospheric CO_2 levels would be 200 ppm higher. The development of calcium carbonate shells by marine organisms, that are fed carbon by phytoplankton, fixes carbonate ions at the same time. These shells sink, taking carbonate from the surface waters and lowering their alkalinity. This process works against the primary biological pump effect and tends to raise pCO_2 in the ocean. The proportion between these two processes influences the total impact of biological activity on the sea's surface pCO_2, and consequently, the CO_2 exchange rate between the air and the sea naturally. The biological pump systems that are ultimately regulated by the supply of nutrients from deep water are thought to be superimposed by the absorption of anthropogenic CO_2, which is assumed to be a purely physical process.

By accelerating natural exchange processes, CO_2 is taken up from the atmosphere. A higher pCO_2 air results in a greater mean atmosphere-ocean concentration gradient, which in turn increases oceanic sink activity and decreases source activity. Geographically distinct upwellings of ancient deep water, rich in organic carbon that may have been isolated from the atmosphere for many years, are produced by the spatial structure of marine circulation. The long-term rate of net CO_2 absorption is ultimately constrained by the speed at which oceanic circulation transports currents of deep water to the surface because the air-sea exchange of CO_2 finds a physico-chemical equilibrium swiftly relative to the slow speed of oceanic circulation. Many factors are probably going to have an impact on the ability of oceans to absorb CO_2 in the future. The main restriction is chemical: as the ratio of bicarbonate to carbonate ions rises, so does the atmospheric CO_2 concentration. The capacity for dissolved CO_2 to dissociate ionically is impaired due to the reduced availability of carbonate ions, which in turn limits the capacity for CO_2 to dissolve from the air into the water in the first place. At greater ambient CO_2 concentrations, this impact is substantial and severely restricts absorption. The net rate of CO_2 absorption is further constrained by the velocity of oceanic circulation, and the reduced solubility of CO_2 in water at higher temperatures would further slow transfer to the oceans. Generally speaking, models predict that the annual atmosphere-ocean CO_2 flux will increase throughout the course of the 21st century, reaching 4.5-6.7 PgC/ yr by the end of the century, but that the pace of increase will slow, especially after the first 30–50 years.

The regional variability of carbon-transfer processes in the terrestrial biosphere makes it difficult to directly assess the terrestrial carbon sink(Keenan & Williams, 2018). Forests are excellent candidates to house the majority of this carbon sink due to their spatial extent, biomass, and productivity. The results of field observations appear to support this. A significant increase in the carbon store in northern forest biomass, of the order of 0.6 Pg/yr, has been suggested by extensive inventories of forest biomass in temperate and boreal regions. Most tropical forest regions lack the resources to conduct such thorough inventories, but a compilation of data from forest plots in forests reveals that these woods are growing in biomass, creating a terrestrial carbon sink of 0.85–0.25 PgC/yr. Estimates from the forest inventory do not account for changes in soil and litter carbon. When forest productivity and biomass rise, it is likely that soil and litter carbon reserves increase as well. If soil and biomass residence times are similar, the increase in soil and litter carbon reserves will be about equal to the increase in reserves in live biomass. This shows that forests account for the majority of the terrestrial carbon sink, but it is likely that savannahs and grasslands also play a role. The restoration of forests on former agricultural fields comes first.

5.8 FERTILIZATION EFFECT

The growing atmospheric CO_2 concentrations are promoting plant photosynthesis, which has an impact on the quantity of carbon stored in plant biomass, litter, and soil-organic carbon, making CO_2 fertilization another possible contributing factor. The degree to which the fertilization effect is constrained by the availability of other nutrients, such as nitrogen and phosphorus, is a significant area of ambiguity. According to some theories, plants may react by directing a larger amount of their additional carbon toward the development of roots and root exudates in order to boost their nutrient supply. All forest ecosystems may be impacted by the high CO_2 effects, although tropical places with naturally high production may see a greater impact. In areas that experience seasonal dryness, increased atmospheric CO_2 also leads to improved water use efficiency, which may expand the growing season of plants. Human activities have improved the global nitrogen cycle by releasing nitrogen oxides during the combustion of fossil fuels and biomass, as well as ammonia with the use of fertilizers in agriculture and industry. Evidence suggests that nitrogen deposition is promoting forest growth in temperate regions; however, this effect is less significant in boreal and tropical regions, which are farther from nitrogen sources and, in the case of the latter, constrained by phosphorus deficiency.

Whatever the reasons for the carbon buildup in forest areas, the carbon balance varies greatly from year to year. This is clear from studies of the regional distribution of atmospheric CO_2 as well as ecophysiological models of forest processes, which show that tropical regions are the main source of much of this interannual variability. Because of increased fire activity (both man-made and natural), reduced photosynthesis, and dry weather in most of the Amazon and tropical Asia during El Nino (Figure 5.4) years, tropical regions become a net carbon source(Liu et al., 2017).

On a longer time scale, it is probable that the terrestrial carbon sink will become less significant as the amplitude of the CO_2 fertilization effect declines and structural limitations restrict the amount of extra carbon that can be stored in forest biomass. It is still unknown what the main limiting variables will be and when they will start to matter. Additionally, it is probable that temperature change may hasten the decomposition of soil organic matter and increase plant respiration, making natural ecosystems net carbon sources. In tundras and high latitude boreal forests, where climatic warming is most pronounced, measurements of CO_2 fluxes indicate that the enhanced carbon sink in biomass is being counterbalanced by the release of soil carbon brought on by the thawing of biomass. The biosphere isn't likely to be a purely passive sink for carbon; instead, it is more likely that the changes affecting the terrestrial carbon sink will result in significant adjustments to the ecological balance of ecosystems, which will have an impact on both species diversity and ecosystem function. For instance, at the most fundamental level, the CO_2 response is significantly higher in plants with a C_3 photosynthetic mechanism than it is in those with a C_4

mechanism. Laboratory studies have shown that responsiveness to high CO_2 varies between species.

At least temporarily, anthropogenic CO2 emissions are expected to increase. There is now almost no doubt about how this will affect the climate. One of the major difficulties of the twenty-first century will probably be figuring out how to reduce these emissions and deal with their effects. It will be necessary to implement a number of initiatives, such as switching to renewable energy sources, improving carbon usage efficiency, and possibly sequestering CO2 in deep sediments or the deep ocean. Preventing deforestation, reducing carbon loss from forests by switching from conventional to reduced-impact logging, and controlling other disturbances like fire and pest outbreaks are some options for biosphere management. Other options include reforestation/afforestation of abandoned or degraded lands, sequestration in agricultural soils through changes to tilling practices, and increased use of biofuels to replace fossil fuels. Human activities had caused the release of approximately 190 PgC from the biosphere by the year 2000. Therefore, an equal quantity of carbon would be stored if all agricultural and degraded lands were to be restored to their former plant cover. The IPCC Third Assessment Report confirms earlier estimates that the potential avoidance and removal of carbon emissions that could be achieved through the implementation of an aggressive program of changing forestry practices over the next 50 years is 60-87 PgC. The tropics could produce about 80% of this quantity. Changes in agricultural techniques may add another 20 to 30 Pg of carbon sink during the same time span, bringing the total carbon sink from land management to 80 to 120 Pg, with a mean annual sink of about 2 PgC/yr(Post & Kwon, 2000).

For instance, undamaged, well-managed forests will nearly always hold more carbon than degraded forests or agricultural fields, despite the fact that the natural carbon sink may diminish with global warming and possibly turn into a net source. Better carbon-focused forest management will therefore almost always lead to net carbon sequestration. According to the IPCC's "business as usual" scenario, land use change and the burning of fossil fuels would emit around 1400 PgC during the course of the 21st century. For the 21st century, emissions are predicted to range between 1800 PgC and 2100 PgC, according to more current and precise emission predictions. Regardless of the scenario's specifics, it is evident that even a comprehensive land carbon-sink program could only offset a small portion of the likely anthropogenic carbon emissions over the next century. The maximum feasible human-induced forest-carbon sink may be exceeded by

5–10 times by fossil fuel emissions alone over the 21st century. It is obvious that absorbing carbon in trees cannot fix the issue of global warming on its own. However, forest-carbon absorption can be beneficial when it forms a substantial part of a set of CO_2 mitigation methods and acts as a temporary carbon sink while other mitigation technologies are being developed. Early-century carbon absorption has a higher impact on lowering temperatures by the end of the century than late-century carbon absorption.

The relative potential of forest-carbon sinks decreases as the century goes on and the size of the necessary carbon reductions grows. Between 10 and 30 percent of the necessary offsets could only be produced by a forest-carbon balance program if it were established in 2050. Therefore, for a program to be effective, the majority of the carbon must be absorbed within the next several decades. Any plan for reducing carbon emissions over the coming decades may benefit significantly from the controlled absorption of carbon in forests. Such a plan could be seen as partially reversing the detrimental climatic and biological effects of earlier centuries' worth of forest clearing. Forest-based carbon absorption cannot be seen as a long-term solution to the global warming issue because a forest's relative potential contribution falls later in time. It can only serve as a temporary fix. Because it controls the amount of carbon dioxide (CO_2), the most significant greenhouse gas (GHG), in the atmosphere, the global carbon cycle is acknowledged as one of the major biogeochemical cycles. Because they store significant amounts of carbon in soil and vegetation, exchange carbon with the atmosphere through photosynthesis and respiration, and are sources of atmospheric carbon when disturbed by human activity or natural events, forests play a significant role in the global carbon cycle. Through modifications to forest land use and management, humans have the power to modify the size of forest carbon stocks and the direction of forest carbon fluxes, changing the forest functions in the carbon cycle. The Kyoto Protocol included land use change and forestry (LUCF) activities since it was understood that they might operate as both generators and sinks of carbon.

The Protocol contains numerous sections that provide for net changes in GHG emissions by sources and removals by sinks on the land resulting from specifically human-induced activities in regard to a country's reduction objective. Afforestation, reforestation, and deforestation since the base year of 1990 are the only things covered by Article 3.3(Perrin, Temmerman, & Laitat, 2000)requiring co-operation between regional and federal administrations. Around a fifth of the total area of Belgium is covered by forests. Around 80% of the productive forests are in the Walloon region. Reported values for

LUCF categories give a potential of 2,057 kt eq. CO2.y-1. Given the existing Regional Forest Inventories (RFI. Since 1990, Article 3.4 has allowed for new human-caused activities such revegetation, agricultural management, grazing land management, and forest management. Emission-allowance trading is permitted between industrialized nations under Article 6. Projects that reduce sources of GHG emissions or improve sinks of GHGs in any area of the economy can result in emission reduction units, provided that these emission reduction units are in addition to any that would otherwise occur. Similar to Article 11, the Clean Development Mechanism (CDM) helps developing nations achieve sustainable development by allowing for the trading of carbon offsets between industrialized and developing nations. In addition to any emissions reductions that would have occurred without the project, those coming from such project activities must be substantial, quantitative, and long-term benefits relevant to climate change mitigation. The issue surrounding the use of biological methods to absorb or reduce carbon emissions led to the decision to restrict LUCF activities under the CDM. In this context, the term carbon sinks refer to direct human-induced changes in the use and management of forests and agricultural areas; it does not relate to enhanced carbon storage that could result from changes in atmospheric CO_2 or increased nitrogen deposition, for instance.

By avoiding deforestation, changing harvesting practices, and switching from conventional to reduced-impact logging, land-use change and forestry activities can reduce carbon emissions; by increasing the area and/or carbon density of forests; and by expanding the storage of carbon in forest ecosystems. In the pilot stage of activities implemented jointly (AIJ), which was established under the Berlin Mandate in 1995, several projects that comprise the LUCF activities previously mentioned have been created. Even in the absence of climate change considerations, the majority of the land-use and forestry techniques discussed above that reduce GHG emissions make sound social, economic, and ecological sense. Because it appears that GHG mitigation is the primary objective of the project, land-use and forestry operations for mitigation are frequently questioned. Instead, LUCF activities can achieve the more traditional goals of managing forests, such as: sustainable forest development; industrial production of wood and fuel; traditional forest uses; protection of soil, water, and biodiversity; recreation; rehabilitation of damaged lands, etc.

Since photosynthesis has been demonstrated to increase at high CO_2 concentrations, it is assumed that plants will absorb more CO_2 in a future atmosphere with an enriched carbon dioxide concentration. Because of the CO_2-fertilization effect, adopting LUCF activities to reduce carbon emissions may lead to carbon gains(Swinglandet al., 2002)Prance, Saunders, Brown, et al., 2002. The Kyoto Protocol is clear that carbon credits must be quantifiable, transparent, and verifiable and that they can only result from direct human-induced changes in the use and management of the land. The IPCC Third Assessment Report confirms earlier estimates that the potential avoidance and removal of carbon emissions that could be achieved through the implementation of a vigorous program of changing forestry practices on roughly 700 million hectares over the next 50 years is approximately 60-87 Gt. The Third Assessment Report also reaffirmed that the least expensive GHG mitigation method involves activities in tropical forest areas. The cost-benefit of tropical forestry over forestry in OECD nations, renewable energy, energy efficiency, and fuel switching becomes more important as total reduction levels rise. Such cost-effective policies in poor nations are more likely to persuade wealthy nations to fund similar initiatives without experiencing significant economic disruptions.

5.9 HANDLING CARBON CREDITS

Typically, deforestation is conducted by burning a sizable amount of biomass. Nitrous oxide, another potent greenhouse gas with the potential to warm the planet more than CO_2, is produced when biomass is burned. Therefore, preventing deforestation also prevents significant nitrous oxide emissions, the amount of which is unknown. It has recently been demonstrated that a large number of mature tropical forests continue to absorb carbon from the atmosphere. Conserving threatened tropical forests prevents GHG emissions and sequesters more carbon, albeit it is unclear what the mechanism for this sink is and whether it is human or not. Changes in forest harvesting have the potential to cut carbon emissions, but the size of these potential savings has not yet been determined. In the 1990s, around 11.5 Mha of tropical forests were harvested annually, according to a recent evaluation of tropical forests by the Food and Agriculture Organization. If reduced-impact logging were to become widespread in the tropics and equivalent savings were assumed, an initial estimate of the prevented carbon emissions would be up to about 20 Mt per year. The extent of carbon emissions prevented would be less with reduced-impact logging since less timber is produced per unit area, which could lead to an increase in the area under logging to maintain timber output. There are ways to address all of these concerns about carbon sinks that either already exist or may be developed using existing principles and technology.

Around the world, there are several LUCF projects in various phases of design and execution, including community forestry, agroforestry, forestation (afforestation, reforestation, and restoration), and forest protection(Cacho, Marshall, & Milne, 2003)Marshall, & Milne, 2003. While the majority of other project types are in tropical nations, the majority of forestation projects are in non-tropical nations. These projects have accumulated a lot of knowledge in the fields of permanence, carbon monitoring, without project baseline development, and leakage prevention up to this point. The lifespans of forests are measured in millennia, and they must survive any alternative technology-based installation as well as any currently advanced target period for reducing atmospheric CO_2. Recent studies have shown that tropical forests continue to store CO_2 for the duration of their existence. Projects are considered permanent for all pertinent purposes if they continue to store or sequester carbon for the chosen goal period or longer. The risk that the stored carbon could be released back into the atmosphere by natural or anthropogenic events such as the failure to enforce contracts, non-compliance with guarantees, uncertain property rights, policy changes, land tenure, and market risks gives rise to concerns about the permanence of biological sinks. The possible influence of future climate change on forest-carbon budgets has recently drawn attention. According to simulations, warming will increase soil and plant respiration, which will reduce a forest's capacity to store carbon. However, a recent analysis of this matter came to the conclusion that there is a lot of uncertainty in these estimates. For instance, the models relied on assumptions such as the assumption of a pristine terrestrial biosphere, the relationship between temperature and soil respiration, which has recently been called into question, business-as-usual CO_2 emission scenarios without consideration for emission mitigation, and no provision for changes in how humans manage the terrestrial biosphere.

The issue of the possibility of atmospheric release of carbon stored in biological systems has been addressed using a number of workable solutions. One is to acknowledge that carbon sinks are a temporary means of reducing GHG emissions and to evaluate the advantages of temporary storage for the environment and the economy. The postponing of climate change, the buying of time for the development and discovery of alternative technologies to reduce emissions, the buying of time for capital stock turnover, the provision of limited, insurable periods, and the provision of a means for host countries to maintain sovereignty and the opportunity to pursue other future development pathways, are some of the economic and environmental reasons. There have been proposals that essentially treat forestry initiatives as nature's rentable service providers. Where permanence is neither assured nor desired, the conventional method of a rental contract for short-term use of an asset is perfect for the transfer of carbon credits. The limited-term carbon credits are for the advantage of the renter (or buyer), but the seller keeps long-term control over the resource. According to these recommendations, the renter would have to replace the credits at the conclusion of the rental period by renting new credits, buying permanent credits, or incurring a debit. The land would be released from all obligations after the credits run out, or the owner might choose to extend the project for an additional time frame and be free to renegotiate.

There is concern that many carbon-sink initiatives would have taken place anyway for business or political reasons unrelated to the commitments related to combating climate change, adding little to the effort to do so. Large-scale exotic tree plantations for pulpwood or sawtimber are an example of this(Engelmark et al., 2001). Although these plantations have the potential to store a significant amount of carbon, they might have been built regardless of climate change due to the substantial financial benefits. Given that the area of industrial plantations in developing countries has increased at a rate of approximately 4 Mha per year over the past ten years or so, tree plantation initiatives for timber products may face significant challenges in demonstrating additionality. However, forestry projects that provide little to no financial gain but yet incur continuous costs for a developer are unlikely to be started at random and are therefore likely to satisfy the additionality requirement. Once additionality has been established, each mitigation project needs to have a baseline or prediction of the carbon emissions or storage under business as usual. To quantify the amount of carbon stored or emitted based on the traditional pattern of land use and forestry, baselines can be developed by extrapolating previous patterns and present conditions. Practically all infrastructure and long-term capital investments, like power plants, are planned using comparable baselines. Other techniques include benchmarking models that are comparable to those used in business and benchmarks for minimum performance.

5.10 LAND USE CHANGES

Changes in land use can be predicted based on a variety of biophysical characteristics, the presence of transportation networks, market accessibility, and agroclimatic appropriateness. Remote sensing data, spatial land-use change models, forest growth models, and field measurements are just a few of the techniques that have already been used to create very straightforward but reliable baselines by project type and region. Leakage is defined as the unplanned reduction or augmentation of GHG benefits brought about by project activities outside of the project boundary. Potential leakage arises from two factors: market effects, which occur when project activities alter the equilibrium between supply and demand; and activity shifting, which occurs when an activity that produces carbon emissions inside a project's boundaries is relocated outside of it.

Figure 5.5. Biodiversity conservation.

One of the most difficult technological problems associated with the creation of carbon projects is the identification and measurement of leakage. It appears to be equally difficult for both land-use and energy initiatives, which has been the topic of research. The sheer possibility of leakage does not automatically render a project unappealing; rather, solutions must be established to either account for or minimize it. The absence of data and short time periods since project initiation have confined experience to a small number of projects. It may be necessary to further improve qualitative techniques as well as initiatives to produce more accurate data at the appropriate level of definition. Forest-based carbon projects can assist host nations and local communities in a variety of extra environmental and socioeconomic ways. Carbon sink projects can provide the funding required to assist nations in achieving numerous local and national sustainable development goals, including technology transfer, the expansion of national parks, the reduction of rural poverty, and improved national capacity for climate change adaptation and, consequently, reduced vulnerability.

The conservation of current biodiversity or its restoration, (Figure 5.5) the preservation of habitats, and the safeguarding of soil and water resources are all examples of environmental co-benefits from forest operations. Improved water flow and quality from protecting water resources reduce siltation and flood risk, which in turn safeguards downstream water users, fisheries, coastal coral reefs, and hydropower infrastructure. The majority of projects involving improving forest management give local landowners significant socioeconomic and environmental advantages. Fuel-wood plantations can be established to lessen the effects on native forests, particularly in arid areas, which may help to slow the rate of desertification. However, biodiversity would be diminished if exotic forest plants were employed in place of natural habitats like grasslands or woods. The claim that there may be less technological transfer of emission-reducing technology to poor nations is one of the uncertainties regarding permitting forestry projects in the CDM. The concern is that carbon sinks will act as a stand-in for clean energy technology transfer to underdeveloped nations or as a deterrent. The advantages of technology transfer, however, will rely on the technical characteristics. For instance, compared to importing large-scale advanced technology, small-scale, locally reproducible technology is expected to spread more quickly and have larger job and poverty reduction effects. Numerous forestry-related innovations, including tree selection, nursery management, enhanced silvicultural techniques, improved forest management and harvesting methods, protection of biodiversity, wildlife management, etc., are amenable to local adoption and diffusion to the rural poor. Projects that include alternate fuel supply would also transfer technologies for biofuel or small-scale electricity(Amigun, Sigamoney, & von Blottnitz, 2008).

Improved crop production technology would be transferred through projects that adopt better agricultural methods to offset the loss of agricultural land. The Kyoto Protocol has been the focus of recent global warming efforts. The employment of biological techniques to absorb or reduce CO_2 emissions has come under scrutiny in that context. Since minimal new technology is required to deploy carbon-sink solutions, they can be readily adopted locally and disseminated to the rural poor. According to the IPCC's Third Assessment Report, biological sinks can contribute significantly, even if only slightly, to the shift to an environment with fewer emissions at a significantly lower cost than other mitigation strategies. Regardless of whether the Kyoto

Protocol is ratified, this is the situation. Well-planned forestry initiatives can help host nations and regional populations in substantial socioeconomic and environmental ways, particularly in the tropics. Regarding the difficulties of regulation, there is no discernible difference between carbon sinks and other mitigation strategies. Utilizing current ideas and methods, a regulatory framework for such initiatives can be put into place. A well-crafted regulatory framework would support sustainable development while also advancing efforts to combat climate change. It would adhere to international accords on biodiversity, desertification, wetlands, and indigenous peoples' rights. Initiatives to protect the environment and the financial markets came together in the final decade of the 20th century. Market-based strategies, like emissions trading, are now widely recognized as a practical way to improve the environment.

5.11 ENVIRONMENTAL MARKETS

Environmental markets are evolving in a manner akin to that of other well-established sectors. The equities, commodities, and fixed-income markets can be used as examples. Trading greenhouse gas emissions (GHG-emissions) is becoming more widely accepted by the corporate sector and governments as a cost-effective method to lessen the risk that human-induced climate change is posing. This is because societies are moving toward a world where carbon is restricted. So, in the case of carbon dioxide (CO_2) emissions, we are seeing the same evolutionary process. The field of conservation can also profit from employing market processes like emissions trading. In order to maintain and restore habitats, carbon sinks like trees can make a significant contribution to the preservation of biodiversity and the improvement of water quality. We can anticipate that in the near future, the financial performance of a firm and its environmental management will be more closely correlated, and market-based methods will be used more to solve concerns like water quality and fisheries. The adoption of a property-like instrument, such as emissions allowances and offsets, offers a mechanism that can guarantee efficient use of the resource and generates a price in a market that wasn't before accessible.

CHAPTER 6

Electrochemical Reduction

CONTENTS

6.1 INTRODUCTION

With 35 gigatons of anthropogenic CO_2 produced by burning or using fossil fuels, the challenge of converting carbon dioxide is significant (Hunt, Sin, Marriott, & Clark, 2010). According to data released by the National Oceanic and Atmospheric Administration of the US Department of Commerce, the average concentration of CO_2 in the atmosphere had increased to 403.38 ppm by September 2017. Serious effects are anticipated on global warming, climate patterns, sea levels, biodiversity, food production, human displacement, and other factors. The most appealing but also technically difficult option is to emulate natural photosynthesis, even though CO_2 sequestration has also been suggested. Links between catalytic, electrochemical, and photo-electrochemical conversion of ambient CO_2 can be built in the context of artificial photosynthesis (Figure 6.1).

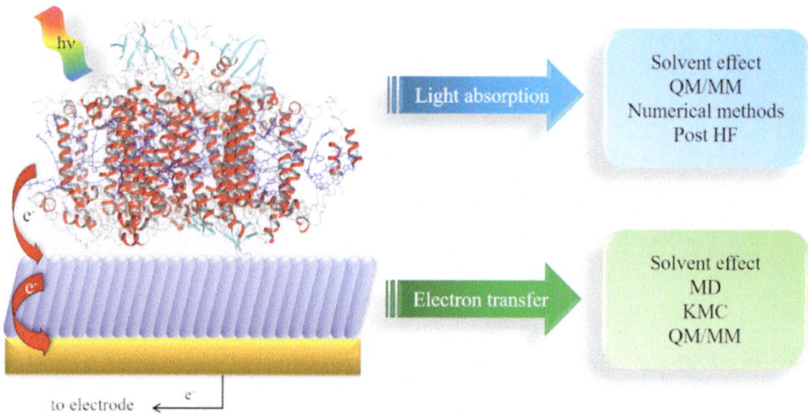

Figure 6.1. Artificial photosynthesis. (Source: Osella S. Artificial Photosynthesis: Is Computation Ready for the Challenge Ahead? *Nanomaterials*. 2021; 11(2):299. https://doi.org/10.3390/nano11020299)

By incorporating functional nanoscale components that cantrap solar light, accumulate CO_2, and selectively reduce CO_2 into products, new technologies that can be extensively used for local carbon emission reduction can be developed. Additionally, there are integration strategies for semi-artificial photosynthesis, such as using PS_1 and PS_2 structures that have been removed from cells and immobilized on electrode surfaces. Even though these nano-integrated systems have the potential to have synergistic benefits, their level of structural complexity currently rarely results in high efficiency or stability. No technological advancements have been made that make it possible to clearly answer this question as of yet. One of the major obstacles to a low carbon economy, according to a sizable community of scientists, is a direct and selective pathway of CO_2 to a valuable carbon structure. Investigating the characteristics of the catalytic sites using electrochemical techniques yields incredibly useful mechanistic information, regardless of the strategy utilized to provide the driving force for CO_2 reduction.

It has frequently been stated that the high symmetry of the carbon dioxide molecule is a major factor in the activation barrier, with a cost in energy frequently incurred by providing a high overpotential (excess thermodynamic reduction potential). Recent research suggests that the binding energies of intermediate species in multi-electron transfer events correlate linearly, leading to high overpotentials for a range of metals. There have also been reports on the thermodynamic statistics for organic media relating to aqueous media. Only molten salts and high temperatures are capable of reducing to elemental carbon, and even nano-carbon compounds. Light-absorbing units (ruthenium bipyridyl photosensitizer), a metal oxide as an electron-accepting moiety, and carbon monoxide dehydrogenase, serves as an example of photoexcitation as the most appealing method to produce highly energetic electrons capable of driving CO_2 reduction. The photo-absorber, which locally produces charge carriers, is a crucial part of the overall process (Jacobsson et al, 2013). In addition to molecular dyes (ruthenium bipyridyl), there are also additional substances like quantum dots, oxide nanoparticles, carbon dots, and heterostructures made of carbon.

6.2 CATALYTIC PROCESS

Synthetic biology and metabolically engineered microbes, together with biological systems, have been generating a lot of interest in catalytic centers. Other outstanding examples of nanoscale hybrid integrated systems are the immobilization of dehydrogenase enzymes or PS_1 and PS_2 centers that have been isolated from cells. The catalytic process can benefit greatly from selectivity. At semiconductor materials like $BiVO_4$, for instance, direct and selective catalytic transformations have been accomplished, resulting in high efficiency in the production of single products like methanol or ethanol. It has been documented that very complex compounds such as lactate or dimethyl carbonate can be generated in a single step. The production of electrochemical reactors for continuous flow CO_2 photo-reduction is another exciting field of research. Catalytic electrodes are typically put in a reactor in contact with an electrolyte solution that is saturated with CO_2, which may be an aqueous, organic, or ionic liquid. Particularly in aqueous media, the solubility of CO_2 in these systems is an important consideration. Under high pressure, the low CO_2 solubility in water can be somewhat improved. To know the solubility restrictions, membrane reactor designs with efficient triple phase border reaction zones can be used. Increasing the number of active sites or altering the material's intrinsic activity are two common ways to modify the activity of catalysts. However, completely new catalysts and structures for catalysts are also required. The intrinsic activity of late transition metals has been thoroughly studied from both an experimental and theoretical point of view.

There are two different categories of catalysts: those that generate format (Pb, Hg, In, Sn, Cd) and those that form CO (Cu, Au, Ag, Zn, Pd, Ga, Ni, and Pt). Due to its tendency to produce a variety of CO_2 reduction products and to encourage carbon-carbon coupling, copper occupies a unique position. Although this classification refers to the primary produced products, the overpotentials needed to drive the CO_2 reduction at these metals differ greatly. Thorough investigations at single crystal electrodes have substantially increased our understanding of the CO_2 reactivity at model surfaces. Using sophisticated *in situ* spectroscopic techniques, new insights are conceivable. On the other hand, metal oxides and other materials with highly correlated electrons are far less well known. By adding alloys or metastable catalytic materials, one can also alter the intrinsic activity of catalysts. Studies on CO_2 reduction at ultrathin Pd layers grown on Au nanoparticles have revealed intriguing details about the impact of the electrical and strain effects. The effective strain rises from less than 1 to 3.5 percent in Pd lattices when the Pd shell thickness is reduced from 10 to 1 nm(Jiang et al., 2015). Rare earth elements and other transition-metal cations are important for the function of molecular catalysts. The ligand structure also has a significant impact on these systems' reactivity. For instance, binding to the ligand rather than the metal center is how zinc porphyrinato complexes catalyze the conversion of CO_2 to CO. The electrolyte salt can be a significant factor. It has also been demonstrated that alkali cations have an impact on the CO_2 reduction product distribution. At copper surfaces, electro-reduction of CO_2 has been seen to be more effective when halide anions are present. The effects of electrolyte cations in ionic liquid media are quite exciting since they have a significant impact on the double layer's structure and make it possible to stabilize reaction intermediates like CO_2 radical anions. Investigating the area between bulk and molecular catalyst materials is fascinating.

6.3 SURFACE AREA

At the moment, nanoparticles which are known to be very effective catalysts but are frequently poorly understood in terms of size distribution and surface chemistry dominate this. The exception is single crystal nanoparticles. Increased active surface area is essential, but the electron effects of shrinking to sizes smaller than 5 nm in diameter are frequently significant as well. In catalysis, the structure of nanoparticles is frequently crucial, and new catalysts offer further potential.

High active surface area microporous catalysts and metal-organic framework catalysts have been proposed. Novel cluster systems, such as those built on the Au, have been proposed and demonstrated to be very effective in converting CO_2. With the development of new nano-architectures, the chemistry of heterogeneous and homogeneous catalysis is becoming increasingly intertwined. By limiting homogeneous catalysts to interfaces, new strategies are possible that combine the practicality of heterogeneous systems immobilized at electrode surfaces with a deeper understanding of the ability to tune molecular systems. Although conducting polymers and immobilized coordinating polymers have been used successfully in prior techniques, there is a growing focus on nanoscale architecture. Cu_2O and surface-immobilized Re metal complexes are mixed with active substrate materials to significantly alter the catalytic reaction pathway. Similar to this, it has been reported that the immobilization of a molecular Mn catalyst on TiO_2 enables stable photo-electro-reduction of CO_2 to CO when used with a photo-anode that produces oxygen(Ješić et al, 2021) (Ješić, Lašič Jurković, Pohar, Suhadolnik, & Likozar, 2021). For life to continue on Earth, biological carbon dioxide reduction is an essential process. Nature has created amazingly ingenious processes for CO_2 reduction and conversion into the most sophisticated and high-energy carbon compounds needed for metabolism and growth. CO_2 is the main chemical building block for these materials. At rates that allow complex organisms to function, enzymes commonly known as carbon dioxide reductases (CO_2Red) efficiently catalyze the reduction of CO_2 into a variety of lower oxidation state carbon compounds. The basic chemical energy source for all non-autotrophic life forms, autotrophs use CO_2 as a substrate and are driven by solar radiation in the case of phototrophs or from reducing chemicals in the case of chemotrophs.

These CO_2Reds operate in a very complex environment. Complex electron transport pathways provide the reducing energy, and regulatory mechanisms continuously produce or degrade enzymes in response to changing environmental conditions. One of the many chemical processes in the metabolic pathways that are involved in cell growth and biomass production is CO_2 reduction. These include systems like the Wood-Ljungdahl pathway and the Calvin cycle and involve a wide range of various enzymes and cofactors. The development of integrated bio-electrochemical CO_2 reduction devices has been fuelled by the potential benefit of using such organisms or some of their components for anthropogenic application. The effective catalysis of specific reactions has been accomplished in electrochemical cells by the incorporation of a single kind of enzyme. But this strategy frequently has issues with high isolation costs, limited turnover, and weak resistance to degradation. As the organism regulates and maintains its internal components, coupling whole organisms into an electrochemical cell offers gains in stability. Autotrophs can also synthesize short C-chain products and/or biomass, however with a lower energy efficiency than single enzymes.

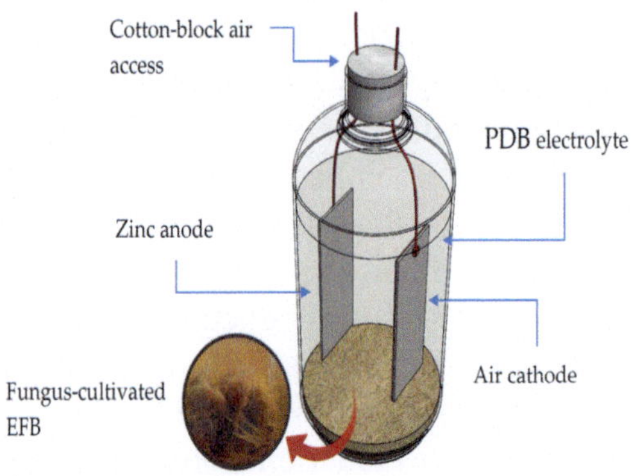

Figure 6.2. Bioelectrochemical cells.

The creation of non-biological catalysis was motivated by the highly optimized blueprint that natural catalysis offers. An example of a CO_2 lowering system whose design is influenced by biological CO_2 reduction in natural photosynthesis is artificial photosynthesis. It is possible to distinguish between the majority of bio-electrochemical systems based on whether the catalyst engages with the electrode directly or indirectly. DET takes place between the electrode and the immobilized biocatalyst(Milton, Wang, Knoche, & Minteer, 2016). Here, the careful choice of the biocatalyst and the way in which it is connected to the electrode are extremely crucial. The NADH cofactor, which can operate as a hydride donor or terminal electron donor, a reduced molecule like H_2 or formate, or a physiologically acceptable redox mediator like methyl viologen are all examples of systems that use soluble reducing equivalents. Therefore, in addition to the catalyst itself, consideration must be given to the interaction, regeneration, and stability of this species.

6.4 BIOELECTROCHEMICAL CELLS

When constructing bioelectrochemical cells, (Figure 6.2) which nearly entirely function in solution, a number of practical factors should be taken into account. One-compartment cells require consideration of factors such as product oxidation at non-selective anodes, whereas two-compartment cells require consideration of factors such as pH gradients and resistance across separating membranes. It may be necessary to operate the cell anaerobically in some conditions, and it is frequently preferable to reduce cell volume to aid in the detection of products at low concentrations. Rate optimization can be achieved by temperature management. The reaction solution must be acceptable for the catalyst and any mediators, and choosing parameters like pH may require making trade-offs between rate and stability, as well as having an impact on the solubility of $CO_2/CO_3H/CO_3$ and the equilibrium position. Consideration should also be given to species produced electrochemically, for instance at practical working potentials. Care must be made to prevent the accumulation of inhibitory products in enzymatic cells. The immobilization of the bio-catalyst is essential in cells using DET, and a number of techniques, including electrostatic control and covalent control, can be used to optimize the binding orientation for optimal activity. Eliminating contamination should be a goal when utilizing online or offline methods to identify CO_2 reduction products, and this can be done in a number of ways. Low molecular weight products from such systems have frequently been detected using chromatographic methods including gas chromatography (GC) and high-performance liquid chromatography (HPLC), as well as other approaches like spectroscopic enzymatic assays.

There are many performance measurements for these systems. The turnover frequency and turnover number reflect the energetic efficiency and selectivity, whereas the overpotential and faradaic efficiency (FE), also known as catalytic selectivity, indicate the energetic efficiency and selectivity. Direct product detection should be used to assist activity metrics. The stability of the system over the short term and long term is crucial for assessing its feasibility. The substrate CO_2 slowly equilibrates with the hydrated bicarbonate form at physiological pH levels. Some carboxylases require bicarbonate as a substrate, while others, including formate dehydrogenase (FDH) and carbon monoxide dehydrogenase, require the non-hydrated form. Carbonic anhydrase (CA), which is found in organisms, may replenish the bicarbonate reservoir of CO_2 quickly by reversibly converting CO_2 with the hydrated form at very fast rates. The rates of CO_2-dependent reactions thus go up overall. CA has been added to enzymatic bio-electrochemical systems that use FDH to boost system activity. Bicarbonate has been proposed as the substrate, although this interpretation frequently fails to take into account the CA-catalyzed quick interconversion of the two forms. Many findings support the hypothesis that CO_2 serves as the substrate for FDH, indicating that the function of CA in these systems is to replenish CO_2.

It has proven successful to use dehydrogenases such as FDH (formate oxidation), glutamate dehydrogenase (glutamic acid oxidation), glutamate dehydrogenase (glutamic acid oxidation), phosphite dehydrogenase (phosphite oxidation), and others. While good for activity and mechanistic research, these systems add another level of complexity to the system, which also needs to be optimized for pH, temperature, and concentration. They also require the regular addition of non-interfering substrates for continued catalytic efficacy. Regenerating NADH using an electrochemical device that either uses direct electron transfer (DET) or a mediated or catalyzed reducing process is an alternative method(Yuan, Kummer, & Minteer, 2019). To achieve low overpotential, high activity, and high specificity, this regeneration must be optimized. Electrochemical reduction of NADH typically occurs after two single electron transfer steps. When NAD1 is first reduced by one electron, a radical intermediate is formed. This radical intermediate can either be further reduced to NADH via proton-coupled electron transfer, which is a slow process, or it can dimerize to form NAD2, which is resistant to enzymatic oxidation. Activity tests should discriminate between these forms, and regeneration systems must be unique to the active NADH form. High overpotentials have traditionally been required for NADH regeneration at metal electrodes through DET, favoring the inactive NAD2 dimer. On carbon-Pt and carbon-Ni electrodes, however, where the metal provided a high concentration of active adsorbed hydrogen, NADH selectivity up to 100% was achieved, optimizing the rather slow proton-coupled electron transfer process.

By adding catalysts or mediators to the system, the regeneration overpotential can also be reduced. Some of the developed transition metal complexes based on Ru, Co and Rh can regenerate NADH. Although their interactions with enzymes have not yet been studied, NAD analogs that are more stable and capable of regeneration than NAD itself have also been produced. As catalysts for CO_2 reduction, enzymes are attractive because they display extremely high energy efficiency for specific products. This is a result of both their exact reaction paths and their capacity to catalyze reactions that are very close to the thermodynamic limit. Many enzymes can function as CO_2 reductases in nature, either by directly reducing CO_2 to produce tiny molecules or by performing the first step in the biosynthesis of complex compounds. These include the carbonate-reducing enzymes FDH, CODH, and ribulose-1,5-bisphosphate carboxylase oxygenase (RuBisCO). Examples of CODHs and FDHs from various organisms catalyze the process in both directions, typically with a kinetic bias for CO_2 or formate oxidation. However, the equilibrium can be changed to favor CO_2 reduction by chemically altering the enzyme through substrate and cofactor concentrations or electrochemically through substrate concentration and potential. The expression or overexpression of native or recombinant enzymes by a host organism, followed by extraction and purification methods, can be used to produce enzymes biochemically. The scale at which a target enzyme may be generated is largely influenced by the need for enzyme-specific methods and the wide range of production efficiency. The use of enzyme-based catalysts for medium- to large-scale industrial CO_2 reduction currently faces a number of significant obstacles that are related to different combinations of poor stability, high O_2 sensitivity, high cost, production difficulties, and low rates, depending on the specific enzyme employed. Enzymatic systems are no longer considered a viable option for industrial CO_2 reduction, despite some success having been had in developing non-reductive CO_2 conversion systems based on the highly active enzyme(Rasouli, Nguyen, & Iliuta, 2022). The design of synthetic catalysts, however, can be influenced by electrochemical investigations, which are useful for comprehending enzyme operations.

6.5 ELECTROCATALYSIS

Electrocatalysis use this intermediate frequently to compete with the hydrogen evolution reaction (HER), resulting in low product selectivity and limiting the FE. It is preferable to use catalysts with low activation energy requirements and numerous electron transfer routes. Two electron routes are preferred in CO_2 reduction methods that involve enzymes. This is typically caused by a cofactor-coupled process involving hydride transfer or a bimetallic reaction center with both a nucleophilic and electrophilic region, respectively. Enzymes can function extremely close to the thermodynamic limit while yet maintaining good specificity, according to electrochemical experiments. Using the Fischer-Tropsch process, carbon monoxide is easily transformed into liquid fuels. It also serves as a crucial feedstock for the carbonylation reaction, which produces a variety of low molecular weight carbon compounds. In the industrial setting, CO is frequently formed at high temperatures, for instance, by the Boudouard reaction. A wide range of metallic electrode materials can also catalyze the two-electron reduction of CO_2 to CO, with control over the metal-CO adsorption strength being a key element in regulating the reaction rate. Numerous species use carbon monoxide dehydrogenases or CODHs to catalyze the reversible conversion of CO_2 to CO. There are two kinds that come from either aerobic or anaerobic bacteria. These come from aerobic bacteria like *Oligotrophacarboxidovoran*, and are typically found to be stable with a Mo and Cu-rich active site. An active site containing Ni and Fe species is found in CODHs from anaerobic bacteria, including *Moorellathermoacetica* (also known as *Clostridium thermoaceticum*), *Rhodospirillum rubrum*, and *Carboxydothermushydrogenoformans* (ChCODH). Although CODH from *Methanothrixsoehngenii* has shown O_2 stability and CO oxidation activity, these are often not stable. This reduction is the initial step in the carbon fixation process in autotrophic organisms using mechanisms like the Wood-Ljungdahl route(Berg, 2011).

For the CO_2 reduction process, CODH and CODH/ACS have both been proposed as potential bio-electrocatalysts. By using X-ray crystallography, the structures of CODHs from each of the aforementioned anaerobic bacteria were identified, and they are identical. There are two [Fe_4S4] B-clusters, which participate in both internal and external electron transport inside the enzyme. The two monomeric units are connected by a unique [Fe_4S4] D-cluster, but its function in structure or electron transmission is not yet fully understood. The C-cluster has received extensive research because it is the active site for the reversible catalysis$_2$. A distorted cubane $NiFe_3S_4$ cluster with an attached unique Fe that is not a part of the distorted cubane makes up this cluster. Due to its existence in the crystal structure of several CODHs, the significance of a bridge sulfide between Ni and unique Fe was contested for a number of years. However, numerous studies have since demonstrated that this is absent from crystal structures that are critical for catalysis, and it is now generally acknowledged that this species is not crucial for catalysis. While protons may be transferred by a transfer network, it is believed that CO_2 or CO substrates will enter the active site by a hydrophobic channel. It is interesting to note that it is been suggested that this may display proton translocation directionality that varies based on the C-oxidation state.

Although functionality is still lacking, there has been tremendous progress in the creation of synthetic catalysts based on the NiFe-CODH C-cluster. From EPR investigations conducted at various applied potentials, four electronically unique forms of the active site C-cluster have been identified. These four forms are known as Cox, Cred1, Cint, and Cred2. However, it is believed that only the two reduced states are involved in catalysis. In the process of CO oxidation, the nucleophilic Ni and electrophilic Fe sites of the hydroxylated Cred1 state are initially bound by carbon in CO and oxygen in water or hydroxide, respectively. If water is bound, water deprotonation may be followed by a change in the CO carbon position, and the hydroxide species is then able to attack the carbon species along with proton loss. The subsequent two-electron transfer results in the loss of CO_2 and attachment of water/hydroxide, resulting in the more reduced Cred2 state. By using the enzyme to transport electrons, the Cred1 state is restored. The distortion from linearity of Ni-bound CO_2 is extremely comparable to that of CO_2, according to studies on CO_2 reduction, and the relatively short Ni-C bond shows significant bonding is taking place. It is believed that the CO_2 reduction reaction is activated by this metal to ligand charge transfer. The electrical contact with the enzyme in bio-electrocatalytic systems is crucial and can take place through mediated or direct electron transfer at an electrode surface. RrCODH and MtCODH/ACS are two CODHs from different organisms that have been shown to mediate CO_2 reduction. The one electron reducer methyl viologen (MV), which has been employed to regenerate CODH during CO_2 reduction, is typically found to be the best suited for high turnover. Many investigations have demonstrated that the presence of CO inhibits CO_2 reduction and that the removal of CO from the solution increases ongoing activity.

When electrochemical mediator regeneration is used, CO_2 reduction typically takes place close to the thermodynamic potential and works best at slightly acidic pH. When employing CtCODH/ACS CO_2, CO_2 was lowered at the thermodynamic limit as opposed to the overpotential of about 100 mV required by the MtCODH/ACS system(Can, Armstrong, & Ragsdale, 2014). Over the course of 24 hours, this turnover drastically dropped once more, most likely as a result of declines in enzyme activity. In 2007, researchers showed that mono-functional Ni-CODH from the anaerobic thermophile *Carboxydothermushydrogenoformans* may be used for DET assays. The tremendous energy efficiency of the system was demonstrated by the fact that only extremely small overpotentials were needed to drive the reaction in either direction. The setup was discovered to be stable for a number of hours, and reducing pH increased relative CO_2 reduction/CO oxidation activity. The solvent-exposed Fe_4S_4 D-cluster was assumed to be the route by which DET to the enzyme occurs. The significant CO_2 reduction activity of this CODH has been attributed to the its negative potential. Michaelis constants have been electrochemically determined in order to gain insights into CODH behavior employing DET electrochemical techniques, also known as protein film electrochemistry (PFE). A better knowledge of the various C-cluster states is also made possible by these insights on the potential-dependent behavior of inhibitors.

6.6 FORMIC ACID

Formic acid is a chemical of significant industrial importance; in 2012, its market value was estimated to be \$450 million. It is employed on a vast scale in the manufacturing of leather and in a range of other industries. It has lately been thought of as an H_2 storage material in terms of energy vectors since it can be easily formed by CO_2 reduction with the addition of two protons and subsequently oxidized in a fuel cell. It serves as a precursor in the methanol synthesis process, which has the potential to be used as fuel. Formic acid and carbon dioxide are interconverted reversibly by the action of Formate dehydrogenases (FDHs). There are numerous FDH structures that can be categorized as either metal dependent or metal-independent. Depending on their metabolic function, these variations exhibit a variety of distinct activities for HCOO oxidation and CO_2 reduction, exhibiting a preferred reaction direction. Aerobic bacteria and methylotrophic yeast both contain metal-independent (MI) FDHs, which typically have the same homodimeric substructure. They range in molecular weight from 70 to 100 kDa and do not contain any metallic or other prosthetic groups. The majority are discovered to be stable, can display activity over a broad pH range and are rather simple to make in large quantities. Some of them, including that from *Candida boidinii* (CbFDH), are marketed. However, the majority exhibit lower activity than metal dependent FDHs and frequently exhibit more activity for formate oxidation than for CO_2 reduction.

By hydride transfer, reversible interconversion between CO_2 and formate occurs and NADH/NAD1 is required. Both the open and closed conformations of the active site, which contains binding sites for both NADH/NAD1 and CO_2/formate, have been observed. The transition from the flexible open, NAD1-unbound form to the more rigid closed, NAD1-bound form after NAD1 binding aids in stabilizing the substrate and cofactor in the ideal location for hydride transfer. Two pores in the structure allow formate or CO_2 to enter the active site in the closed form. Small variations in these structures between FDH variants have been connected to variations in CO_2 reduction activity. For this enzyme class, formate oxidation has received more attention than CO_2 reduction. The C in formate and the C4 position in NAD1, undergo a direct two electron hydride transfer during this reaction, and this transfer is the reaction's limiting step. Atomic-scale crystallographic studies have helped to establish that NAD1 assumes a so-called bipolar form in which the C4 has improved positive charge and electrophilicity, facilitating oxidation when bound in the active site of FDH from *Moraxella* sp. Formate generating electrolysis cells, which use MI-FDH, typically incorporate commercial FDH and NADH in solution in a one or two-compartment bio-electrochemical reactor. MI-FDH has been frequently used in electrochemical CO_2 reduction systems because it is relatively easy to obtain and has desirable handling properties. In such systems, indirect NADH regeneration has been accomplished via a [Cp*Rh(bpy)Cl]Cl complex (where Cp* is pentamethylcyclopentadienyl and bpy is bipyridine), itself regenerated at a copper cathode, and in one example via the redox dye neutral red, itself regenerated at a graphite cathode. It was discovered that the rate of formate production increased when the potential was moved negatively due to advancements in NADH regeneration. However, FDH is inhibited at NADH values higher than 0.3 mM. Using *Saccharomyces cerevisiae* FDH and long chain viologen species immobilized on FTO particles, a mediated/DET intermediate system incorporating both viologen mediators and immobilized enzymes was created(Liao, Liu, & Meng, 2022). When a nine-carbon, carboxylate-terminated viologen alkyl linker was utilized, formate was generated over the course of three hours at a constant rate of 7.6 mmol/h.

To increase activity, some practical measures can be done, such as pH optimization and cautious enzyme selection. For instance, in a system containing FDH from *Phaseolus aureus*, CO_2 reduction activity rose 20 times when the pH was lowered from 8.0 to 6.3. Several structure-function relationships have been proposed to explain these differences. Metal-dependent (MD) FDHs are often more complex than MI-FDHs, and there are numerous variants in terms of metal cluster content and quaternary structure. These enzymes range in size from 80 to 510 kDa and are often membrane-bound, cytoplasmic, or periplasmic, depending on their intended use.

Although membrane-bound FDHs have also demonstrated CO_2 reduction activity, those that naturally act as CO_2 reductases have been discovered in the periplasm or cytoplasm of various examples of prokaryotes. Most iron-sulfur clusters in metal dependent FDHs are necessary for electron transport within the enzyme, and they are reduced via electron transfer at a donor/acceptor site close to the protein. One or two pyranopterin guanine dinucleotide cofactors, a SeCys or Cys, and S coordinate Mo or W in the active site, which is mostly conserved between variations. Two more His and Arg residues are found to be directly engaged in catalysis and are likewise highly conserved. Additional active site modifications make it easier for the formate C-H bond to be oxidatively broken, which has been demonstrated by isotopic substitution studies to be rate limiting. After that, CO_2 is released from the enzyme and exits most likely through a hydrophobic channel. Changing the active site SeCys to Cys in the *E. coli* FDH-H reducedkcat by a factor of 300, which may be related to the different pKa values of the two residues and subsequent protonation differences. Typically, mechanistic studies have emphasized CO_2 as the product, and it is unclear if CO_2 reduction will actually reduce CO_2. By way of iron-sulfur clusters, electrons in MD-FDH go from acceptor species close to the enzyme surface to the active site. Artificial electron donor/acceptor molecules like viologens can interact with these terminal acceptor units. These enzymes have substantially quicker CO_2 reduction rates than MI-FDHs. (Figure 6.3)

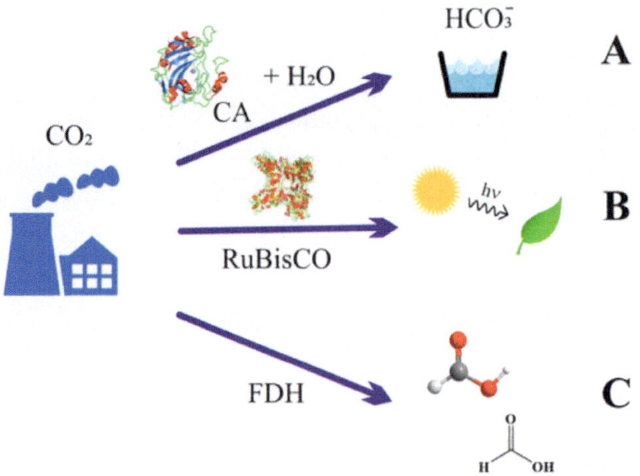

Figure 6.3. Three pathways to incorporating CO_2 into soluble carbon forms by using enzymes. (A) Formation of hydrogen carbonate using carbonic anhydrase; (B) photosynthesis performed using RuBisCO protein; (C) reduction to achieve C_1 in reduced form, formic acid, catalyzed using FDHs.

Syntrophobacterfumaroxidanswas used to extract W-FDH1, which included at least nine iron-sulfur clusters, while *Escherichia coli* was used to make Mo-FDH-H, which only contained one iron-sulfur cluster. Both of these samples were obtained under anaerobic conditions. Both enzymes have been shown to reversibly catalyze CO_2 reduction when immobilized onto an electrode in anaerobic conditions with high FE(Contaldo et al., 2021). This was related to either a molecular version of the species that is inactive in solutions or a distinct iron-sulfur cluster. The hydrogen-dependent CO_2 reductase enzyme was isolated from the acetogen *Acetobacteriumwoodii* using a slightly different method. Through electron transfer subunits, this four-subunit protein connects an iron-containing hydrogenase with a Mo-FDH. Instead of using H_2 to drive CO_2 reduction, it was also conceivable to employ reduced ferredoxin, which may be produced by CO oxidation at a CODH from *A. woodii*. This enzyme has not yet been included in a system that regenerates reducing equivalents electrochemically. The NAD-dependent, metal dependent FDHs are a subclass of metal-containing FDHs. These active sites contain metals, but they typically need NAD to be present as a terminal electron acceptor. It is interesting to note that the oxygen-tolerant, cytoplasmic NAD-dependent Mo-FDH from *Rhodobactercapsulatus* can oxidize HCOO. Its active site is known to have a cysteine rather than a SeCys ligand, indicating that SeCys is not a need for catalysis even though mechanistic details have not yet been clarified. The metal-dependent FDHs are typically found to be quite oxygen sensitive, which restricts their usefulness. This has been connected to the metal-coordinating SeCys residue; while some Se-free W-FDHs have been demonstrated to be stable, it has also been demonstrated that in some instances, replacing SeCys with Cys can result in a significant loss of catalytic activity. Many NAD-dependent, metal-dependent FDHs have demonstrated significantly greater oxygen tolerance. Although formate synthesis was not demonstrated and catalysis was slow, NAD/W-FDH from *Clostridium carboxydivorans* exhibits persistent NADH oxidation activity under aerobic conditions for several days. Cys is incorporated in place of SeCys in the active site of cytoplasmic NAD/Mo-RcFDH, which is discovered to be oxygen tolerant.

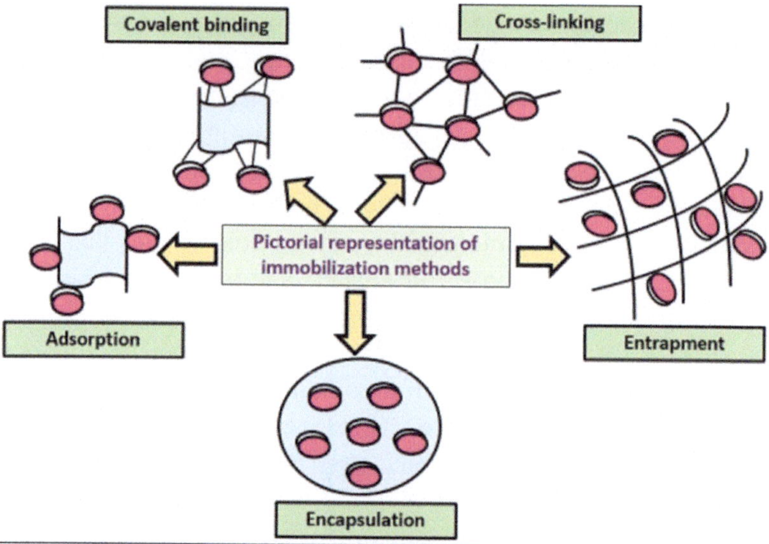

Figure 6.4: Enzyme immobilization. (Source:)

6.7 ENZYME IMMOBILIZATION

Immobilization (Figure 6.4) and bio-encapsulation have been recognized as crucial methods for stabilizing and improving such systems. The enzymes are somewhat confined as a result, which can stabilize them and stop them from unfolding while allowing some degree of dynamic motion. Although large diffusion rates to the bulk solution are still required in multi-enzymatic systems, the close proximity of enzymes also raises the local intermediate substrate concentration. It is crucial to utilize enzyme ratios based on the respective native activities of the enzymes for optimal efficiency. The activity of the enzymes was also improved by optimizing the pore size in these silica gel matrices. It was demonstrated that enzyme-containing titania microparticles, produced by a biomimetic protamine-mediated solution process, increased MeOH yield from 10 percent to 50 percent over the course of an 8-hour reaction. The majority of the improvement was shown in the final four hours of the measurement, which suggests an increase in the enzyme half-life in the reaction conditions. These capsules had a two-layered shell with a silica outer layer with 3.6 nm pores and an inner layer of catechol-modified gelatin (GelC), which allowed for good substrate, cofactor, and product diffusion. With FDH immobilized in the silica layer, FaldDH immobilized in the GelC layer, and ADH immobilized in the capsule core, precise control over the relative position and loading of the various enzymes was obtained. The strong particles could also be recycled and utilized again after being stored. The fact that neither system's measured change in NADH conversion differed from the other is significant since it suggests that the encapsulation favors the complete reduction of CO_2 through to MeOH. Intriguingly, the conversion rate started very quickly in the free enzyme system before it degraded, whereas a maximum rate was seen in the capsule system after 4 hours. This can be ascribed to variations in substrate and product local concentrations and mobility within the two systems.

Cofactor regeneration was accomplished biochemically by oxidizing glutamic acid and catalyzing glutamate dehydrogenase(Greschner et al., 2014). Although immobilization lowered the cofactor's activity by half, it was still able to interact with the enzymes. The manufacturing activity of MeOH was greatly boosted by immobilizing the enzyme into silica nanocapsules, as was demonstrated for earlier systems, by up to 55 times that of the free enzyme. Microbes play a crucial role in the food chain, the creation of oxygen, the decomposition of dead materials, symbiotic connections and other aspects of the Earth's ecosystem. Microbes are interesting in terms of constructing strong self-sustaining systems for controlled CO_2 reduction due to a number of fundamental traits that make them living organisms, such as the capacity for growth and reproduction. The Calvin cycle describes how CO_2 is converted into distinct carbon-based chemicals in autotrophs like photosynthetic microorganisms like cyanobacteria and algae to produce biomass. But certain organisms, including methanogens and acetogens, use CO_2 as an electron acceptor during anaerobic respiration and produce short-chain compounds like methane or acetate as a by-product of ATP synthesis. One microbial CO_2 reduction pathway that produces single carbon byproducts is methanogenesis, and the catalytic cycle for this process has been thoroughly characterized. The process begins with the attachment of CO_2 to methanofuran, and formylmethanofuran dehydrogenase then catalyzes a two-electron reduction to produce formylmethanofuran.

A number of species, including terminal c-type cytochromes and hydrogenases, have been connected to cathode electron transport. Interactions through mediators like viologens have been studied. For a number of applications and reactions, including fermentation control, understanding fermentation processes, promoting cell growth, microbial fuel cells, and electro-biosynthesis, electrochemical technologies have been integrated with microbial systems. These systems typically contain microorganisms that have been cultured in a growth media solution and use a mediator to transport electrons to the bacterium. *Geobactersulfurreducens* and other bacteria that are capable of directly transferring electrons to an electrode have been used in mediator-less systems, while there are many other microbes as well. Microbes are normally cultivated in an electrolyte media solution over a period of several hours to days. One strategy for promoting CO_2 conversion through such organisms is cathodic H_2 production. However, the four-electron transfer anodic oxygen evolution reaction (OER), which is kinetically constrained, makes water splitting for electrochemical H_2 synthesis a low-efficiency process that is only practical with extremely active catalysts. In order to avoid using H_2, several MBEC systems have either used DET as an electrode or redox mediators.

6.8 MICROBIAL SYSTEMS

Microbial systems often stay active for a long period as long as they receive the right nutrients, which is one of their specific advantages. After the cell had been operating for three months, healthy biofilms were seen. The Z-scheme of photosynthesis describes how energy is transferred from photons to reducing equivalents and then used to power reduction activities in plants(Yun et al., 2011). The photosystems I and II (PSI and PSII) protein complexes, which are attached to the thylakoid membrane, are where photons are first absorbed and excited states are formed. An electron is promptly transferred from the PSII reaction center to the acceptor pheophytin in the excited state of PSII after passing through a series of species with progressively higher positive redox potentials that ultimately connect to PSI. At PSII, this transport chain makes charge separation easier and reduces electron recombination. Electrons produced by water oxidation, which is catalyzed at the PSII manganese center, are used to replenish oxidized PSII. Excited PSI moves an electron from chlorophyll A_0 to ferredoxin-NADP oxidoreductase, which then reduces NAD1 to NADH, via nearby redox species. The electron transfer from PSII regenerates oxidized PSI. The transport chain between PSII and PSI promotes proton translocation across the thylakoid membrane when an electron passes through cytochrome, resulting in proton gradients that support the synthesis of ATP. These bio-inspired artificial photosynthesis systems use one or two photocatalysts and follow an electron transport pathway that is comparable to a condensed Z-scheme. Because these are regenerable through anodic water oxidation, no species are needed. Systems that have a catalytic cell that is indirectly coupled to the light absorber are taken into consideration.

CHAPTER 7

Post-Combustion Absorption of Carbon

CONTENTS

7.1 INTRODUCTION

Loss of biodiversity, (Figure 7.1) an increase in extreme weather events, reduced agricultural production, ocean acidification, and a decrease in freshwater availability are just a few of the many negative effects that climate change is anticipated to have on the local and global environment.

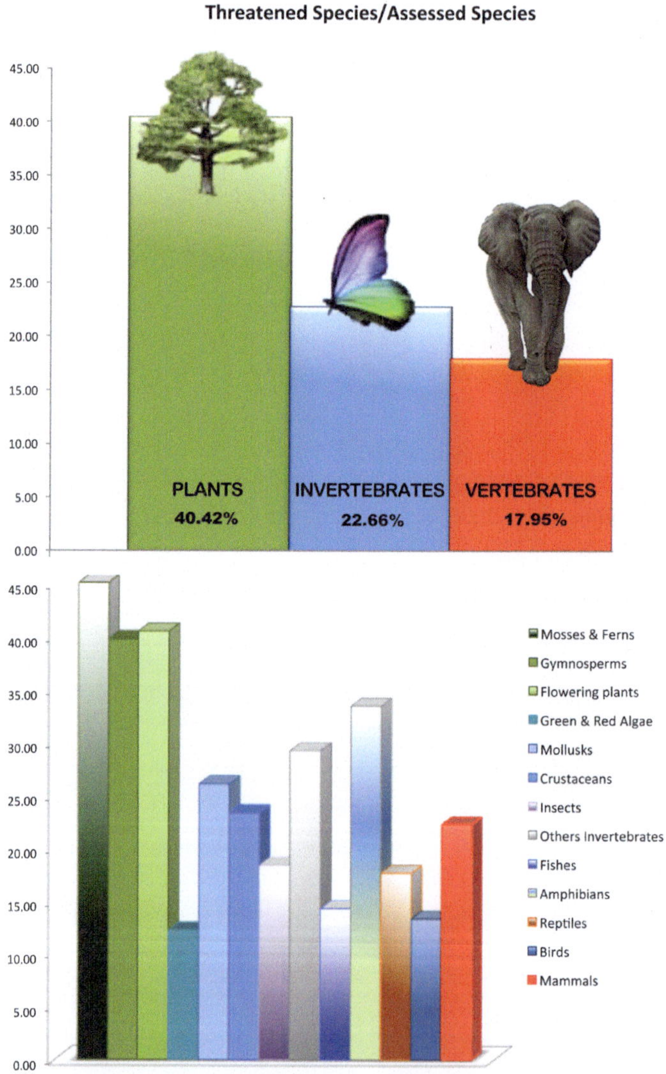

Figure 7.1. Loss of biodiversity. (Source: Palombo MR. Thinking about the Biodiversity Loss in This Changing World. *Geosciences*. 2021; 11(9):370. https://doi.org/10.3390/geosciences11090370)

Industrial emissions of greenhouse gases like CO_2, CH_4, N_2O, and fluorinated gases are linked to human-induced global climate change. The primary source of greenhouse gas emissions worldwide is the combustion of fossil fuels and industrial activities, which produces CO_2(Quadrelli& Peterson, 2007). It is quite likely that these impacts, along with those of other human causes, were the primary reason for the warming that has been observed since the middle of the 20th century. Population growth (P), per capita economic activity (G/P), energy intensity (E/G), and carbon intensity of energy consumption (F/E) all affect CO_2 emissions in different ways. Enhancing industrial energy effectiveness will not only lower their energy intensity but also help businesses save money by using fewer resources. Investments in more energy-efficient processes will pay off in conventional economic activities, but lowering the carbon intensity of energy generation does not directly produce these advantages. Therefore, unless they offer a more affordable alternative, additional steps are needed to encourage the introduction of energy carriers with reduced carbon footprints. Large-scale wind turbines and solar cells are just two examples of the renewable energy conversion technologies that have advanced significantly during the past 25 years. Hydropower and nuclear energy conversion, both of which are regarded as inexpensive electricity sources, have also continued to advance. Despite these advancements, fossil fuels continue to be a major component of many existing energy conversion systems, including those used to generate electricity and transport people and products. Decarbonizing an energy system that is essentially carbon-based is obviously a huge problem.

7.2 EMISSIONS

Global coal use has increased by 73% since 2000, and in 2013, it was responsible for 44% of all CO_2 emissions. At current consumption rates, the world's recoverable coal reserves are sufficient for 120 years. Therefore, it is anticipated that these reserves will be needed to support a large portion of the continued growth in China and India. The utilization of fossil fuels like coal, oil, and natural gas is made possible through CO_2 capture and storage (CCS), but with substantially fewer CO_2 emissions than normal. The foundation of CCS is the separation of CO_2 at some stage of the industrial or energy conversion process, followed by compression, transport, and long-term storage. Attention was drawn to this method of reducing CO_2 emissions from fossil fuel-fired power plants and industrial operations in a 2005 IPCC special report on CCS. The International Energy Agency (IEA) has also indicated that under the scenario allowing a 2°C global temperature increase by the end of the 21st century, CCS could contribute nearly 6 gigatonnes of CO_2 emission reduction by 2050. This was done through the lowest-cost scenario modeling for a portfolio of low-emission energy technologies. The temperature will rise by 6°C during the same time period under the business-as-usual scenario, and there will be 42 gigatonnes more CO_2 emissions. In other words, CCS would contribute roughly 14% to overall CO_2 emission reductions. Although natural gas and other high-pressure gas streams will be the initial uses for CCS, coal and gas-fired power plants will progressively require its use. The IPCC has determined that if CCS is not implemented, the global costs of combating climate change will increase by 138%. CCS is a financially appealing strategy for combating climate change. This is mostly due to the fact that CCS is currently the only technology capable of considerably reducing emissions from fossil fuel-fired power plants. The demand for energy will remain to rise, driven by population growth and global economic growth. This is especially true in China, and in the subsequent few decades, it is also likely to be the case in other Asian countries, South Africa, and Brazil. Coal in particular, which is abundant in the majority of these nations, offers the most affordable option for industrial-scale power production and other energy-related activities. The flue gas produced when the fuel is burned in conventional power plants is often released into the atmosphere. The simplest way to reduce the emissions of these power plants is to equip them with a post-combustion CO_2 capture (PCC) unit(Hanak, Biliyok, Yeung, & Białecki, 2014)largely through a post-combustion capture (PCC. PCC does not necessitate the introduction of new power plant concepts based on gasification, reforming, or combustion in a nitrogen-depleted atmosphere, nor does it necessitate significant technical changes to the power plant architecture. Such changes or new types of power plants are likely to require significant investment and lengthy development processes that are less aligned with the CO_2 emission reductions required to limit the consequences of climate change. This is due to the scale of power plant equipment.

Standard techniques can be used to make the necessary power plant improvements, however, there may occasionally be physical restrictions on their execution. Because of this, not all power plants can integrate PCC systems. In addition to financial and technological constraints, access to a possible CO_2 storage facility may also prevent its deployment. As a component of a larger CCS chain, PCC has the potential to drastically cut CO_2 emissions from existing power plants both now and in the future. India, with its plentiful coal resources, is anticipated to take a big share of new plants in the next decades. The majority of the more recent plants are in China, South Korea, India, and Japan. Small-scale commercial production of CO_2 from flue gas is used for a variety of purposes. When the liquid absorbent is heated again, CO_2 is released. Only acid gases react making the chemical reaction selective. The application of the reactive amine absorption method was the initial focus of research activities connected to climate change mitigation. A variety of new substances, including piperazine, amino-methylpropanol, aminoethyl-amino-ethanol, amino-methyl-propane-diol, diethyl-amino-ethanol, and blends, have been added to the amine family in recent years. Alternative separation techniques based on membranes, cryogenics, or solid sorbents are also being actively researched. Although they potentially have advantages in terms of cost, simplicity, energy efficiency, and ease of use, they have yet to be tested on a pilot-plant scale. In the past, natural gas, related gas, and industrial gas streams have been treated with acid gas (CO_2, H_2S) using amine-based gas-liquid absorption techniques. This has given us a solid scientific and technological foundation for expanding our operating experience to flue gas treatment. However, the conditions for PCC and acid gas treatment differ significantly from one another. The low partial pressure of carbon dioxide in flue gases will result in a higher energy requirement for the separation process; the continuous presence of oxygen in the feed gas will cause greater chemical action on organic capture agents such as amines; and the gas treated in a PCC process is directly emitted, resulting in higher investment costs relative to the amount of CO_2 separated from the feed gas.

Only if conditions are set that place a value on CO_2 emission reduction can the task of a 6 gigaton CO_2 emission reduction by 2050 be realized, which could need up to 3000 large-scale integrated projects(Holappa, 2020) energy, industry, buildings, transport, and cities". The challenge falls on all sectors, especially energy production and industry. In this regard, the recent progress and future challenges of greenhouse gas emissions and energy supply are first briefly introduced. Then, the current situation of the steel industry is presented. Steel production is predicted to grow by 25–30% by 2050. The dominant iron-making route, blast furnace (BF. Although the initial steps in technology deployment have been accomplished, more must be done to build a legislative, commercial, and community framework that will support the further development of CCS technologies. Post-combustion capture (PCC) will play a crucial role in this procedure. In a broader sense, carbon dioxide capture, use, and storage (CCUS) call for the extraction of CO_2 from a gas mixture and the creation of relatively pure CO_2 that can be used or stored. To lower the atmospheric CO_2 emissions from fossil fuel-fired power plants, PCC uses CCUS on the flue gas. A combination of primarily N_2, H_2O, O_2, and other minor components like NO_x and SO_x needs to be separated from the CO_2. Oxyfuel and chemical looping combustion are two alternatives to reduce CO_2 emissions from power plants. In these scenarios, the fuel is burned while being exposed to highly concentrated oxygen, either in the form of metal oxide or O_2. As a result, the flue gas has a high concentration of CO_2. Physical separation based on variations in boiling temperatures, vapor pressure, or molecule sizes is always an option, although these procedures are frequently time-consuming.

7.3 PCC PROCESS

Carbonic acid, H_2CO_3, is formed when CO_2 combines with water. This acid then mixes with bases in proton exchange reactions to create non-volatile bicarbonate and possibly carbonate ions. The amine(s) in solution can only react with CO_2, and it is these reactions that give aqueous amines their desirable CO_2 absorbing capability and absorption rate. Though conceptually simple, the PCC process must be reversible due to the significant energy requirement for the manufacturing of the chemicals used to absorb the CO_2, meaning that during their production, more CO_2 is formed than can be absorbed later. CO_2 is absorbed into a relatively cold amine solution that has a high affinity for CO_2, which drives the cyclic process of PCC. Then, the absorbent is heated to change the chemical equilibria and the solubility of CO_2, which reduces its affinity for CO_2. The remaining CO_2 is then emitted in a specified proportion. In a process known as a packed column, the absorbent and the CO_2 gas combination come into contact. This is a column that is packed with a substance that is porous. Stainless steel sheets that have been perforated are typically layered to create a honeycomb structure (structured packing), or stainless steel rings that are a few millimeters in diameter and randomly distributed. The absorbent is added at the top of the column and runs downward over the packing's flat, solid surface. As a result, the gas and liquid come into contact with a lot of surface area. Increasing amounts of CO_2 are absorbed by the absorbent as it descends the column. This is a typical counter current operation in which the fresh absorbent at the top of the column will absorb the majority of the residual CO_2 that is still in the gaseous phase.

The gas phase has a high CO_2 content and will pick up CO_2, even if there is already a significant amount of CO_2 in the absorbent(Aaron & Tsouris, 2005)approximately 7 Gt of carbon are emitted to the earth's atmosphere each year. A large portion of this carbon is in the form of gaseous CO2, and approximately 30% of this CO2 comes from fossil fuel power plants. In addition to rising levels of atmospheric CO2, the earth's temperature is increasing. Since CO2 can act as a trap for heat (similar to the glass in a greenhouse. The absorbent is collected, pumped to the column, and heated once it reaches the base. The absorbent is once more inserted at the top of the packing column used for stripping. At the bottom of the stripping column, steam is formed, which is then used to eject heat into the absorbent and dilute the CO_2 that is discharged. The released CO_2 flows up through the packing interstices while the absorbent runs downward over the packing's solid surface. The PCC process is actually a complicated method that calls for enormous reaction columns for the absorption and desorption processes as well as a lot of energy for pumping, heating, etc. There are definite benefits to collecting CO_2 as rapidly as possible, including shorter absorption times and smaller absorber columns. This reduces the capital expense. Reducing the amount of energy required for the stripping column, which is the stage of the process that requires the greatest energy, has substantial additional benefits. The energy needed to raise the temperature of the absorbent and the energy used by the chemical reactions combine to create the amount of steam needed to dilute CO_2 and maintain a driving force for desorption. Chemical and physical methods are used to remove CO_2 from a gas mixture via reactive chemical absorption. Getting the gas and liquid into close contact and efficiently transferring heat to or away from the absorbent are the main goals of the engineering design.

Once dissolved, CO_2 initiates a chemical reaction to generate a chemically unique product when it comes into contact with an absorbent molecule. This reduces the concentration of dissolved CO_2, allowing more CO_2 (g) to penetrate the gas-liquid contact. The sole distinction between CO_2 absorption and desorption is whether it is entering or leaving the liquid phase. Diffusion is what moves molecules from the liquid and gas to the gas-liquid interface. The spontaneous movement of molecules from one area of higher concentration to another is known as diffusion. It is comparable to spontaneous heat transfer, in which heat moves from a heated zone to a colder region. The temperature, viscosity, and flow regime of the liquid, as well as the intensity and rate of the chemical reactions, have an impact on this diffusive transport. An aqueous amine solution is the absorbent generally employed for CO_2 absorption It comprises an amine (or amines) that are water soluble and can interact chemically with CO_2 in a variety of ways. Amines are employed because they are easily available, because they react with CO_2 and release it when heated, and because they have generally good physical properties. The PCC process, which uses potassium carbonate solutions, is a notable example. Physics here refers to the characteristics that control this movement as well as the net transit of material between phases. In the case of oxygen in seawater, for instance, the oxygen molecules go from being in a highly mobile gaseous environment to being surrounded by far less mobile water molecules when a gas dissolves in water. This state of lowest entropy is attained when there is a concentration gradient between a gas and a liquid and a soluble gas has dispersed itself across the gas and liquid phases to the point where equilibrium is reached and the liquid is saturated at the dominant gas conditions.

The speed at which gas molecules can diffuse through the gas and liquid phases determines how long it takes a gaseous-liquid system to equilibrate. The amount of time it takes to attain equilibrium is greatly reduced if either phase mixes. In comparison to the condensed liquid phase, diffusion occurs more quickly in the gas phase because the molecules are more mobile and have higher kinetic energies. Therefore, the amount of time it takes to attain equilibrium is usually governed by the rate of diffusion in the liquid phase. This is dependent on both the viscosity of the liquid and the molecular volume of the diffusing molecule. Depending on the kinetics of the reactions, the creation of the chemical equilibrium may take some time. If the column is long enough and the gas and absorbent flows are slow enough, the equilibrium will result in very little CO_2 remaining in the gas phase at the top of the column. The absorbent will contain the most absorbed CO_2 near the bottom of the column since this is where the partial pressure of CO_2 in the gas phase is at its highest(Puxty, Rowland, & Attalla, 2010)low energy requirements for CO2 regeneration and resistance to oxidative and thermal degradation. However, due to its small molecular weight and large vapour pressure absorption must be carried at low temperature to minimise ammonia loss. In this work the rate of CO2 absorption into a falling thin film has been measured using a wetted-wall column for aqueous ammonia between 0.6 and 6molL−1, 278–293K and 0–0.8 liquid CO2 loading. The results were compared to 5molL−1 MEA at 303 and 313K. It was found that the overall mass transfer coefficient for aqueous ammonia was at least 1.5–2 times smaller than MEA at the measured temperatures. From determination of the second-order reaction rate constant k2 (915Lmol−1s−1 at 283K. The difference in the CO_2 loading of the absorbent between the maximum loading at the bottom of the absorber and the minimum loading at the bottom of the stripper is the key indicator. Cyclic capacity is the cause of this variation. The more CO_2 thatan amount of absorbent can catch, the higher its cycle capacity. It takes energy to heat the CO_2 -rich absorbent. It is crucial to reduce this energy because it accounts for a sizable portion of the PCC plant's operating costs.

To keep the partial pressure of the released CO_2below the equilibrium CO_2 partial pressure and to maintain a driving force for desorption, it is necessary to dilute the CO_2 throughout the stripping column. The less steam is needed for dilution, the higher the equilibrium partial pressure. This is determined by how much the affinity of the absorbent for CO_2 declines with rising temperature, which is in turn determined by how the position of the chemical equilibrium moves with temperature.

7.4 AMINES USED IN PCC

There aren't many amines with good characterization available for PCC. Specific amines that are preferably chemically diverse are studied, and methods like the quantitative structure activity relationship (QSAR) can be developed to point a developing chemist in the direction of more effective amines. Di-amines are a stoichiometric solution; one amine group of the di-amine reacts with CO_2 while the other picks up the proton. The most prevalent candidate is piperazine, abbreviated PZ. PZ has a number of appealing qualities, including one of the fastest reactions with CO_2, the ability to form the di-carbamate, which accepts two CO_2 molecules, but which also causes the released protons to occupy the other two amine sites, reducing the effective stoichiometry to 1:1. Designing a molecule with two functional moieties, one that reacts quickly and the other that serves as a base with a significant negative enthalpy is an obvious solution. The average chemical reaction time and the diffusion time both affect the enhancement factor. When a gas and liquid come into prolonged contact or when the reverse reaction predominates, these conditions are not met. However, they can be used if the contact time is brief.

For primary, secondary, and tertiary amines, the reverse processes for CO_2 desorption happen relatively quickly. They might even be seen as immediate. The speed at which CO_2 may transition from the liquid phase to the gaseous phase is what determines how quickly desorption can occur. Energy consumption and capital costs have been continuously reduced as a result of advancements in process design and amine choice. The standard absorption/stripping method for removing H_2S and CO_2 from natural gas and other fuel gases serves as the foundation for amine scrubbing, which is used to extract CO_2 from combustion gas. Natural gas sweetening technology was developed in 1930. This method has been used in hundreds of plants to treat hydrogen and natural gas for acid gas using aqueous monoethanolamine (MEA) or other common amines. Blends containing rate-promoting substances like piperazine and tertiary amines like methyldiethanolamine (MDEA) are used in research(Fürhacker, Pressl, & Allabashi, 2003). It is possible to use amines alone or in combination with rate promoters, such as aminomethyl propanol (AMP). MEA and PZ are examples of secondary or primary amines that can be employed alone or in combination with potassium carbonate, amines, or tertiary amines to increase the rate. These amine solutions must contain water in order to function. The water content is tuned to balance viscosity and the inherent CO_2 capacity when there are no additional restrictions present, such as solid precipitation. Viscosity reduction improves mass transport and heat exchanger performance. The CO_2 absorber design has been enhanced over the one used to treat acid gas.

7.5 BYPRODUCTS AND OTHER ISSUES

Water vapor that is condensed without heat recovery as it passes overhead reduces the efficiency of the stripper. To remedy this vulnerability, numerous process configurations have been suggested and put to the test. The stripper's absorption liquid is taken out in the middle and pushed through a second exchanger in series with the hot solution and the primary cross exchanger. This technique recovers a large portion of the latent heat of water from the stripper above, similar to the advanced flash stripper. Since there are no variables for optimization in this design, it is typically less effective than the sophisticated flash stripper. Each of these absorbent qualities is linked to a significant capital cost/energy use trade-off with an advanced process setup. The cross exchanger's capital cost and energy loss depend on its liquid CO_2 absorption capacity. The amount of packing needed for the absorber and the driving power behind it are determined by the rate of CO_2 absorption. The capital cost and lost work for the compressor, reboiler, and stripper are determined by the heat of CO_2 absorption and the maximum working temperature. As an alternative, oxidation intermediates such as organic peroxides may build up through oxidation in the absorber before thermally decomposing at the stripper temperature and causing more oxidation. Since dissolved metals seem to be essential to this mechanism, oxidation would be reduced by avoiding corrosion, applying thermal reclamation, and taking other measures to reduce the buildup of dissolved metals. Reduced duration and temperature under stripper conditions would also reduce the quantity of oxidation. The choice of amine has an impact on the rate of oxidation because of the shuttle mechanism used to switch between the absorber and stripper temperatures. It is possible that oxidation inhibitors won't work with the shuttle mechanism. Several inhibitors were effective under absorber conditions but ineffective during cycling trials.

Practically all amines exhibit formate and ammonia as oxidation byproducts (von Gunten, 2003). The formate equilibrates with the corresponding amide of a parent primary or secondary amine under stripper conditions. In many systems, aldehydes and the corresponding imines and hemiaminals also occur frequently. There are stable aldehyde species that can arise in MEA systems from ammonia, MEA, glyoxal, and formaldehyde, however many of the aldehyde species are in reversible equilibria. One such stable species is hydroxyethyl imidazole (HEI). PZ oxidizes to the stable hemiaminal 2-hydroxy-piperazine. Unfortunately, these systems are not immune to the creation of nitrosamines since primary and tertiary amines can oxidize to produce some secondary amines. MEA undergoes oxidation to produce hydroxyethyl glycine. Diethanolamine is produced in huge amounts when MDEA is oxidized. Under stripper conditions, secondary amines and nitrite combine to form carcinogenic nitrosamines, which may have substantial secondary environmental effects. By reacting with the liquid amine to form oxidized amine and nitrite, residual NO_2 will be eliminated in the amine absorber. The NO_2 will be eliminated by secondary and tertiary amines with a 90–99% efficiency. Primary amines react more slowly and might only eliminate 50%. Less than 20% of the absorbed NO_x undergoes minor side reactions in the absorber that result in the production of nitrosamine. Nitric oxide (NO), in small quantities, will be absorbed together with the NO_2. In the absence of NO_2, NO is not eliminated, and oxidation of the amine is the main source of nitrite, with a yield of less than 10% of the oxidation rate. Nearly all of the nitrite interacts with the amine at the temperature and residence duration of the stripper.

7.6 NITROSAMINE AND OTHER BUILDUPS

Nitrosamine in the proper amount is produced by the reaction with the secondary amine. Primary and tertiary amines slow down the reaction, which also yields oxidation products and N_2. Nitrosamines will still be formed at the stripper conditions because absorption liquids containing tertiary and/or primary amines will have secondary amines that are byproducts of degradation. The nitrosamine thermally degrades into oxidation products and probably N_2O at the stripper temperature. When the rate of generation from NO_2attack into the absorber is equal to the rate of heat degradation, the nitrosamine will build up to a steady state concentration.

Nitrosamine will likewise be destroyed by thermal recovery(Eizember et al, 2002). High pH catalyzes thermal decomposition, therefore thermal recovery at 150°C with the addition of NaOH will create the necessary conditions to remove nitrosamine from the reclaimer waste. Therefore, amine selection, high-temperature stripping, NO_2 removal in the scrubber, and upstream NOx controls can all be used to manage nitrosamine accumulation. There should be little risk of nitrosamine emissions through the water wash if there is little buildup. If specific amines produce volatile secondary amines as degradation products, this could result in unsolvable issues with gas emissions. Otherwise, the risk of leaking liquid inventory containing nitrosamine will be the most important problem connected to the creation of nitrosamines. Although the water wash at the absorber gas exit can control modest levels of amine volatility, severe amine volatility will call for a more expensive water wash system. A highly volatile amine, however, may be used to minimize amine aerosol emissions. Flue gas with substantial aerosol paricles, such as H_2SO_4 may produce amine emissions as liquid aerosol, which are both environmentally and economically problematic. The presence of an amine concentration that is high in CO_2 in the aqueous nucleus is one explanation. The CO_2 loading is lower in the amine solution. Since the amine vapor pressure over the bulk solution is often higher than that of the aerosol with most amines, the amine diffuses into the aerosol, carrying CO_2 and water with it. The nuclei may be caught in the water wash if they expand to a size bigger than 2.5 mm. Results should be anticipated to vary depending on the type of amine solution, method, and flue gas source with this complex mechanism.

It is not permissible to use expensive amine compounds. Even if oxidation and thermal degradation can be reduced, the flue gas will still contain episodic and steady-state impurities, particularly from the combustion of coal, which will lead to amine losses and prohibitive makeup costs. Future generations of enhanced liquid absorbents with higher efficiency and reliable, sustainable capture techniques are still open to development. It is obvious that the development of novel liquid absorbents is subject to shifting benchmarks and environmental regulations. The decision to choose a new absorbent is a multifaceted effort that is complicated by an inherent and inescapable trade-off: new absorbents, while frequently able to successfully exhibit the required improvements in one criterion, sometimes suffer a considerable and negative penalty in another. To gain the greatest performance advantages, the complete capture process must be improved and centered around the liquid absorbent. The integration of the PCC process into the power plant's steam cycle also needs to be taken into account. A robust development and screening process is necessary to give the following generations of liquid absorbents a wide range of properties while allowing for the flexibility to tailor the liquid absorbent and process around integration with the power station. This is done while keeping an eye on evolving benchmarks. Many substances, usually amines and alkanolamines, have been proposed as potential CO_2 capture agents. Regardless of whether a single- or multi-component absorbent blend is preferred, most liquid absorbents can be investigated using the screening methods. In their most basic form, screening studies aim to identify and evaluate important absorbent qualities through a succession of smaller-scale laboratory measurements.

7.7 MOLECULAR MODELING

Molecular modeling and chemical simulation can give chemists in the lab the essential platform to design, and create new absorbents(Marć, Kupka, Wieczorek, & Namieśnik, 2018). The stability constants of carbamates and the protonation constant of the amine, on the other hand, are important determinants of the energy requirement of absorbents; higher efficiency solvents lead to lower liquid circulation rates and lower energy requirements. One must assess and comprehend these fundamental kinetic and equilibrium qualities apart from other chemical and physical processes in order to characterize a single absorbent. It is thus possible to employ the set of relevant chemical constants (kinetic and equilibrium) and thermodynamic constants (Arrhenius, activation energy, enthalpy, and entropies) characterizing these reactions to intelligently direct and assess the viability of developing amine liquid absorbents. The research on the quick and reversible reactions between CO2 (aq) and amines is best done using the analytical approach, which is frequently used for the study of low-reaction kinetics in the millisecond time frame. Due to the inherent simplicity of the flow measurements, the resulting kinetic constants quantifying the formation of carbamates/carbamic acids are thought to be more reliable than kinetic constants ascertained from the interpretation of bulk CO2 absorption measurements. This is true even though actual operations often demand diluted amine solutions since the low physical solubility of CO2 is constrained by the high Henry constant for water. Without the greater quantities needed by other methods and in the absence of commercial quantities of the liquid absorbents, new adsorbents can still be evaluated.

The user has access to a number of detection options to track the development of the reactions. Given that many of the reaction products in amine solutions have an electrical charge (ionic species), conductivity is a common method of detection. Poor conductivities and signal-to-noise ratios at the low reactant concentrations necessary to see the quick reactions between CO_2 and amine solutions are a drawback of conductivity detection. The majority of kineticists favor spectrophotometric detection, which is a reliable option. The method entails keeping an eye out for variations in the absorbance as the reaction develops. The majority of aqueous amine solutions call for a modification to this method because there are no meaningful absorption spectra in the visible range between 350 and 700 nm. In these situations, measuring the absorbance of the solution after adding one or more colored acid-base indicators allows for the indirect observation of the kinetic processes. Thus, a function of time is used to monitor the pH, which is closely related to the kinetic and equilibrium reactions in the solution. A crucial component of this technique is the choice of the acid-base indicator, which must be made in such a way that the colored forms of the indicator are active in the pH range of the kinetic processes. Importantly, a circulating liquid supplied by an external water bath is required to thermostatically regulate the amine and CO_2 samples inside the device. A thermocouple located inside the flow reactor measures the precise temperature of the solution. The kinetic reactions can be carried out under this control at a variety of temperatures, expanding the available data to include activation energies(Marković, Grgur, & Ross, 1997)for the first time, that at a fixed temperature (274 K.

To increase the overall analytical strength, reactant concentrations are frequently changed. When employing dissolved CO_2 (aq) as a reactant, mass flow controllers (MFCs) can be used to change the gas mixture's composition, resulting in a variety of initial reactant concentrations. The subsequent reaction between the carbamate solution and an acid solution containing a colored acid-base indicator that has been chosen to provide the required color change in the corresponding (low) pH zone takes place in the halted flow. The creation of carbonic acid, as well as the formation and breakdown of carbamic acid, causes a change in the absorbance of an indicator, which is tracked. Due to the presence of colored acid-base indicators in the solutions, the traces correlate to variations in observed absorbance that occur during the synthesis and degradation of carbamate/carbamic acid. The analysis of the kinetic measurement data is frequently just as crucial as the actual kinetic data measurement itself. In the past, a variety of analysis techniques have been used to extract the kinetic constants and mechanistic details of the formation and decomposition reactions. Traditionally, pseudo-first-order conditions are used to perform stopped-flow reactions, which makes it much easier to analyze the data under the assumption of finite depletion of one of the reactants. The analysis and comprehension of amine systems have substantially improved in more recent years due to improvements in processing power and the creation of sophisticated modeling software. With this method, the accuracy of kinetic evaluation is greatly increased by combining the analysis of a higher number of individual measurements. The kinetic constants are thus regressed on the full series of measurement data in this analysis as opposed to conventional analysis, which analyzes each measurement separately. The analysis can take into account intricate reaction processes and mixed data sets that go beyond the usual pseudo-first-order reaction conditions to incorporate second-order conditions. The kinetic contribution of the reaction between CO_2 (aq) and hydroxide (OH), which is produced when the amine deprotonates water in an alkaline environment, must also be taken into consideration. The hydroxide reaction can contribute significantly to the overall reported kinetics even at low amine concentrations, hence allowances/corrections for the hydroxide reactionare frequently made.

7.8 THERMODYNAMICS

It is possible to calculate the corresponding thermodynamic parameters (activation energies, enthalpies, and entropies) for the reactions because measurements may be made at a variety of temperatures. The associated activation energy and thermodynamic parameters for the reactions are obtained by further analyzing the rate constants in terms of the Arrhenius, Eyring, and Van't Hoff relationships. The production of carbamates as byproducts of the interaction of CO2 with amines is of general interest to both the fundamental biological and chemistry communities. The equilibrium constant or stability constant is frequently used to describe the reaction along with the corresponding protonation constant of the carbamate to carbamic acid. The rate and extent of the carbamate formation reaction are strongly coupled to the stability of the carbamate(Gebauer, Völkel, & Cölfen, 2008). Due to the need for a second amine molecule to take up the proton released during carbamate production; highly stable carbamates restrict the overall reaction stoichiometry to 0.5 mol CO2 per mole of amine. In comparison, less stable carbamates can attain reaction stoichiometries of up to 1 mol of CO2 for every mole of amine. It is not surprising to see a lot of studies attempting to quantify the chemical behavior in CO2 -loaded amine solutions in an effort to increase the chemical effectiveness of amine absorbents. During the regression of equilibrium constants and speciation data from bulk vapor-liquid equilibrium measurements, any quantitative information on the species composition can and is frequently employed to give extra clarity. A potent method used to investigate the electromagnetic surroundings of the nuclei in small to medium-sized organic compounds and to clarify intricate structural aspects is nuclear magnetic resonance (NMR) spectroscopy. Both the proton 1H and carbon NMR spectra result from close interactions between adjacent atoms in a molecule, giving rise to distinct spectral signals for atoms in distinct chemical environments within the molecules. Similar to this, Nitrogen NMR spectroscopy can be used to investigate molecules with nitrogen atoms in their structures, however, this method is only regarded as beneficial for qualitative verifications and is hence unusual for research in CO2 capture.

7.9 NMR

The analysis of amine and CO_2 solutions benefits greatly from the use of NMR spectroscopy. It is possible to assess the concentrations of the amine and carbamate species, as well as their protonated forms, independently in the presence of CO_2 since amine solutions provide an additional and duplicate set of peaks in the spectra for the carbamate. This naturally implies that there is enough resolution to completely discern the peaks. The positions of spectral peaks along the horizontal axis of an NMR spectrum that are caused by the resonance frequency of a nucleus in relation to a reference molecule, known as the chemical shifts, can also be used to identify the precise atoms or groups that are responsible for the peaks. This knowledge is crucial when dealing with amines that have many reactive amine groups, which produce a large number of peaks that correspond to two carbamates. Fast proton exchange on the NMR timescale, which makes it difficult to discern between the protonated and deprotonated forms of species involved in protonation equilibria, is a minor drawback to using NMR spectroscopy. As a result, all NMR peaks are the total of the two pH-dependent forms, making it difficult to measure the distinct forms without complex deconvolution and data fitting. The preparation of liquid samples needed for NMR analysis can be done *in situ* using a titration-type method or in batches using experiments of this sort(Cunniff et al., 2022). In the first, samples are made by passing a stream of pure CO_2 gas through an amine solution at a specific temperature and throughout a range of CO_2 loadings, which are measured by measuring mass gain/loss or pressure decrease in the CO_2 cylinder. In the latter technique, samples with known CO_2 loadings are formed by adding HCO_3 to the amine solutions in the appropriate proportions. The solutions are subsequently subjected to a series of pH alterations by the addition of acid to mimic the changes in pH conditions and, consequently, speciation, encountered throughout the CO_2 - absorption process. For the evaluation of organic compounds in aqueous solution, the choice of solvent is crucial, and a modification of the conventional NMR approach is needed. Dimethyl sulfoxide (DMSO) or deuterium oxide (D_2O) is typically needed to lock the sample into the instrument's magnetic field before spectra may be obtained. However, the choice of solvent has a considerable impact on the solubility data and equilibrium constants established in the solutions, which become solvent specific, and has a strong influence on speciation. It is usual practice to fix this problem by inserting a small glass capillary into the sample tube together with a reference substance, such as 3-(trimethylsilyl) propionic acid, and an organic solvent to act as the locking agent. To reduce CO_2 loss, the sample tubes are then sealed. To guarantee consistency in the observed spectra, the NMR equipment internally regulates and maintains the temperature of the samples.

Based on the peak integral areas and the known total concentration of the original amine present in the solution, the relative concentrations of the amine and carbamate species are estimated. Although NMR spectroscopy is a potent technology, it is not without flaws. Due to the increased rates of chemical exchange at high temperatures, substantial peak widening is frequently observed, which can seriously impede the quantitative interpretation of NMR spectra. Apart from the change in the chemical shifts of the amine peaks as a function of pH, amines that produce little to no carbamate do not exhibit any relevant $_1H$ NMR spectra.

7.10 OTHER EXPERIMENTAL TECHNIQUES

The measurement of gas-liquid absorption rates has been done using a variety of experimental techniques, each of which has its own inherent advantages and disadvantages that frequently make it difficult to interpret the measurement data and compare the resulting absorption rates among the techniques. Each technique was formed to make it easier to translate measurement data into physical processes and chemical reaction components using well-established theoretical concepts and modeling techniques. This makes it easier to compare the data generated by various procedures and liquid absorbents. The stirred-cell reactor, laminar-jet absorber, and wetted wall column reactor are among the techniques that are readily available and are used most frequently. Here, the wetted-wall column will be thoroughly explored along with the general concepts of these techniques. Due to irregularities in absorption rates caused by wall effects, which can be defined as absorption occurring into the amine liquid present on the inside wall of the absorber column itself, controlled measurements using cutting-edge laboratory equipment are preferred over absorption measurements using packed columns. Liquid materials frequently have a tendency to travel along nonlinear paths and distribute themselves unevenly across the packing(Battat, Weitz, & Whitesides, 2022). The techniques described here are suitable for the clarity needed to grasp at a fundamental level, but they are not applicable at larger scales, where packed columns are often favored because of their significantly greater surface areas. Water flow through the jacket is often used to regulate the temperature of the solution. Alternately, to maintain isothermal conditions, heating/cooling components are frequently added to the amine liquid contained in the flask.

Although the stirred-reactor design seems to be simple by nature, if the cell is not controlled carefully, subsequent interpretations of the measurement data can be time-consuming and frequently inaccurate. The requirement for the development and maintenance of a smooth, flat, and ripple-free liquid surface, while simultaneously maintaining adequate agitation rates that are required to achieve a well-mixed solution, is a significant problem with the operation of the stirred reactor. Adding baffles inside the vessel allows for the use of slower stirring speeds while retaining high mixing efficiency, which is a frequent way to solve this problem. The laminar-jet and wetted-wall column contactors both absorb CO_2 into a flowing amine liquid, which is a comparable process. The methods differ in how the amine solution and gas delivery system are moved. The amine solution is pumped up through the column's interior from the reservoir, emerging via tiny holes at the peak before falling naturally over the outside of the column's surface. A gas stream containing CO_2 and N_2 is introduced to the system at the foot of the column, and it comes into countercurrent contact with the falling liquid there. A closed loop is formed by collecting the amine solution at the bottom of the column and recirculating it into the liquid absorbent reservoir. As a result, in order to prevent the formation of concentration gradients or the considerable buildup of CO_2 in the film, the liquid running through the column is continuously refilled with fresh solution from the reservoir. It's crucial to manage the liquid flow to have a smooth, even coating covering the column surface, assuring complete coverage. Calibrated liquid flow meters are used to measure the total liquid flow rate in the column, which is normally operated between 0.05 and 0.3 L/min. As the solution is returned to the reservoir at the base of the column after absorption, its temperature is routinely measured.

References

1. Aaron, D., & Tsouris, C. (2005). Separation of CO2 from Flue Gas: A Review. Separation Science and Technology, 40(1–3), 321–348. https://doi.org/10.1081/SS-200042244
2. Abedini, A., & Torabi, F. (2014). Oil Recovery Performance of Immiscible and Miscible CO2 Huff-and-Puff Processes. Energy & Fuels, 28(2), 774–784. https://doi.org/10.1021/ef401363b
3. Aerts, R., Verhoeven, J. T. A., & Whigham, D. F. (1999). Plant-Mediated Controls on Nutrient Cycling in Temperate Fens and Bogs. Ecology, 80(7), 2170–2181. https://doi.org/10.1890/0012-9658(1999)080[2170:PMCONC]2.0.CO;2
4. Al-Ghussain, L. (2019). Global warming: Review on driving forces and mitigation. Environmental Progress & Sustainable Energy, 38(1), 13–21. https://doi.org/10.1002/ep.13041
5. Allam, R., Martin, S., Forrest, B., Fetvedt, J., Lu, X., Freed, D., … Manning, J. (2017). Demonstration of the Allam Cycle: An Update on the Development Status of a High Efficiency Supercritical Carbon Dioxide Power Process Employing Full Carbon Capture. Energy Procedia, 114, 5948–5966. https://doi.org/10.1016/j.egypro.2017.03.1731
6. Amigun, B., Sigamoney, R., & von Blottnitz, H. (2008). Commercialisation of biofuel industry in Africa: A review. Renewable and Sustainable Energy Reviews, 12(3), 690–711. https://doi.org/10.1016/j.rser.2006.10.019

7. Aranda, E., Marco-Urrea, E., Caminal, G., Arias, M. E., García-Romera, I., & Guillén, F. (2010). Advanced oxidation of benzene, toluene, ethylbenzene and xylene isomers (BTEX) by Trametes versicolor. Journal of Hazardous Materials, 181(1), 181–186. https://doi.org/10.1016/j.jhazmat.2010.04.114
8. Arcis, H., Coulier, Y., Ballerat-Busserolles, K., Rodier, L., & Coxam, J.-Y. (2014). Enthalpy of Solution of CO2 in Aqueous Solutions of Primary Alkanolamines: A Comparative Study of Hindered and Nonhindered Amine-Based Solvents. Industrial & Engineering Chemistry Research, 53(27), 10876–10885. https://doi.org/10.1021/ie501589j
9. Awuor, C. B., Orindi, V. A., & Ochieng Adwera, A. (2008). Climate change and coastal cities: The case of Mombasa, Kenya. Environment and Urbanization, 20(1), 231–242. https://doi.org/10.1177/0956247808089158
10. Ball, M., & Wietschel, M. (2009). The future of hydrogen – opportunities and challenges. International Journal of Hydrogen Energy, 34(2), 615–627. https://doi.org/10.1016/j.ijhydene.2008.11.014
11. Basset-Mens, C., & van der Werf, H. M. G. (2005). Scenario-based environmental assessment of farming systems: The case of pig production in France. Agriculture, Ecosystems & Environment, 105(1), 127–144. https://doi.org/10.1016/j.agee.2004.05.007
12. Battat, S., Weitz, D. A., & Whitesides, G. M. (2022). Nonlinear Phenomena in Microfluidics. Chemical Reviews, 122(7), 6921–6937. https://doi.org/10.1021/acs.chemrev.1c00985
13. Berg, I. A. (2011). Ecological Aspects of the Distribution of Different Autotrophic CO2 Fixation Pathways. Applied and Environmental Microbiology, 77(6), 1925–1936. https://doi.org/10.1128/AEM.02473-10
14. Berners-Lee, M., Hoolohan, C., Cammack, H., & Hewitt, C. N. (2012). The relative greenhouse gas impacts of realistic dietary choices. Energy Policy, 43, 184–190. https://doi.org/10.1016/j.enpol.2011.12.054
15. Betzold, C., Castro, P., & Weiler, F. (2012). AOSIS in the UNFCCC negotiations: From unity to fragmentation? Climate Policy, 12(5), 591–613. https://doi.org/10.1080/14693062.2012.692205
16. Bodansky, D. (2016). The Paris Climate Change Agreement: A New Hope? American Journal of International Law, 110(2), 288–319. https://doi.org/10.5305/amerjintelaw.110.2.0288

17. Boiral, O., & Heras-Saizarbitoria, I. (2017). Managing Biodiversity Through StakeholderInvolvement: Why, Who, and for What Initiatives? Journal of Business Ethics, 140(3), 403–421. https://doi.org/10.1007/s10551-015-2668-3
18. Bougie, F., & Iliuta, M. C. (2012). Sterically Hindered Amine-Based Absorbents for the Removal of CO2 from Gas Streams. Journal of Chemical & Engineering Data, 57(3), 635–669. https://doi.org/10.1021/je200731v
19. Boyd, E., Corbera, E., & Estrada, M. (2008). UNFCCC negotiations (pre-Kyoto to COP-9): What the process says about the politics of CDM-sinks. International Environmental Agreements: Politics, Law and Economics, 8(2), 95–112. https://doi.org/10.1007/s10784-008-9070-x
20. Boyd, P. W., & Ellwood, M. J. (2010). The biogeochemical cycle of iron in the ocean. Nature Geoscience, 3(10), 675–682. https://doi.org/10.1038/ngeo964
21. Brechet, Y. (2004). Physically Based Models for Industrial Materials: What For? In Continuum Scale Simulation of Engineering Materials (pp. 231–248). John Wiley & Sons, Ltd. https://doi.org/10.1002/3527603786.ch9
22. Breidenich, C., Magraw, D., Rowley, A., & Rubin, J. W. (1998). The Kyoto Protocol to the United Nations Framework Convention on Climate Change. American Journal of International Law, 92(2), 315–331. https://doi.org/10.2307/2998044
23. Brønn, P. S., & Vrioni, A. B. (2001). Corporate social responsibility and cause-related marketing: An overview. International Journal of Advertising, 20(2), 207–222. https://doi.org/10.1080/02650487.2001.11104887
24. Cacho, O. J., Marshall, G. R., & Milne, M. (Eds.). (2003). Smallholder agroforestry projects: Potential for carbon sequestration and poverty alleviation. https://doi.org/10.22004/ag.econ.289093
25. Can, M., Armstrong, F. A., & Ragsdale, S. W. (2014). Structure, Function, and Mechanism of the Nickel Metalloenzymes, CO Dehydrogenase, and Acetyl-CoA Synthase. Chemical Reviews, 114(8), 4149–4174. https://doi.org/10.1021/cr400461p
26. Cannell, M. G. R. (2003). Carbon sequestration and biomass energy offset: Theoretical, potential and achievable capacities globally, in

Europe and the UK. Biomass and Bioenergy, 24(2), 97–116. https://doi.org/10.1016/S0961-9534(02)00103-4

27. Cass, L. (2005). Norm Entrapment and Preference Change: The Evolution of the European Union Position on International Emissions Trading. Global Environmental Politics, 5(2), 38–60. https://doi.org/10.1162/1526380054127736
28. Chichilnisky, G., & Heal, G. (Eds.). (2000). Environmental Markets: Equity and Efficiency. Columbia University Press. https://doi.org/10.7312/chic11588
29. Christiansen, A. C., Arvanitakis, A., Tangen, K., & Hasselknippe, H. (2005). Price determinants in the EU emissions trading scheme. Climate Policy, 5(1), 15–30. https://doi.org/10.1080/14693062.2005.9685538
30. Christiansen, Atle C., & Wettestad, J. (2003). The EU as a frontrunner on greenhouse gas emissions trading: How did it happen and will the EU succeed? Climate Policy, 3(1), 3–18. https://doi.org/10.3763/cpol.2003.0302
31. Churchill, R. R., & Ulfstein, G. (2000). Autonomous Institutional Arrangements in Multilateral Environmental Agreements: A Little-Noticed Phenomenon in International Law. American Journal of International Law, 94(4), 623–659. https://doi.org/10.2307/2589775
32. Clark, B., & York, R. (2005). Carbon metabolism: Global capitalism, climate change, and the biospheric rift. Theory and Society, 34(4), 391–428. https://doi.org/10.1007/s11186-005-1993-4
33. Contaldo, U., Guigliarelli, B., Perard, J., Rinaldi, C., Le Goff, A., & Cavazza, C. (2021). Efficient Electrochemical CO2/CO Interconversion by an Engineered Carbon Monoxide Dehydrogenase on a Gas-Diffusion Carbon Nanotube-Based Bioelectrode. ACS Catalysis, 11(9), 5808–5817. https://doi.org/10.1021/acscatal.0c05437
34. Conway, W., Beyad, Y., Feron, P., Richner, G., & Puxty, G. (2014). CO2 Absorption into Aqueous Amine Blends containing Benzylamine (BZA), Monoethanolamine (MEA), and Sterically Hindered/Tertiary Amines. Energy Procedia, 63, 1835–1841. https://doi.org/10.1016/j.egypro.2014.11.191
35. Cunniff, S. S., Schaef, H. T., Burton, S. D., Walter, E. D., Hoyt, D. W., Loring, J. S., & Bowers, G. M. (2022). Interlayer Cation Polarizability Affects Supercritical Carbon Dioxide Adsorption by Swelling

Clays. Langmuir, 38(50), 15540–15551. https://doi.org/10.1021/acs.langmuir.2c02139

36. D'Amato, G., Chong-Neto, H. J., Monge Ortega, O. P., Vitale, C., Ansotegui, I., Rosario, N., … Annesi-Maesano, I. (2020). The effects of climate change on respiratory allergy and asthma induced by pollen and mold allergens. Allergy, 75(9), 2219–2228. https://doi.org/10.1111/all.14476
37. DeStefano, C. J. (2015). Honey Bee Nutrition: What We're Not Monitoring. Bee World, 92(2), 50–53. https://doi.org/10.1080/0005772X.2015.1118965
38. Durairaj, S. K., Ong, S. K., Nee, A. Y. C., & Tan, R. B. H. (2002). Evaluation of Life Cycle Cost Analysis Methodologies. Corporate Environmental Strategy, 9(1), 30–39. https://doi.org/10.1016/S1066-7938(01)00141-5
39. Eizember, R. F., Vogler, K. R., Souter, R. W., Cannon, W. N., & II, P. M. W. (2002, May 1). Destruction of nitrosamines. Treatment of nitrosamines with various acids and halogens. https://doi.org/10.1021/jo01319a026
40. Ekins, P., & Speck, S. (1999). Competitiveness and Exemptions From Environmental Taxes in Europe. Environmental and Resource Economics, 13(4), 369–396. https://doi.org/10.1023/A:1008230026880
41. Engelmark, O., Sjöberg, K., Andersson, B., Rosvall, O., Ågren, G. I., Baker, W. L., … Sykes, M. T. (2001). Ecological effects and management aspects of an exotic tree species: The case of lodgepole pine in Sweden. Forest Ecology and Management, 141(1), 3–13. https://doi.org/10.1016/S0378-1127(00)00498-9
42. Erisman, J. W., Galloway, J., Seitzinger, S., Bleeker, A., & Butterbach-Bahl, K. (2011). Reactive nitrogen in the environment and its effect on climate change. Current Opinion in Environmental Sustainability, 3(5), 281–290. https://doi.org/10.1016/j.cosust.2011.08.012
43. Feest, A. (2013). The utility of the Streamlining European Biodiversity Indicators 2010 (SEBI 2010). Ecological Indicators, 28, 16–21. https://doi.org/10.1016/j.ecolind.2012.10.015
44. Fenton, P., Gustafsson, S., Ivner, J., & Palm, J. (2016). Stakeholder participation in municipal energy and climate planning – experiences from Sweden. Local Environment, 21(3), 272–289. https://doi.org/10.1080/13549839.2014.946400

45. Firger, D. M., & Gerrard, M. (2010, December 15). Harmonizing Climate Change Policy and International Investment Law: Threats, Challenges and Opportunities [SSRN Scholarly Paper]. Rochester, NY. Retrieved from https://papers.ssrn.com/abstract=1733985
46. Fisher, D. R., Leifeld, P., & Iwaki, Y. (2013). Mapping the ideological networks of American climate politics. Climatic Change, 116(3), 523–545. https://doi.org/10.1007/s10584-012-0512-7
47. Fuchs, S., & Balling, N. (2016). Improving the temperature predictions of subsurface thermal models by using high-quality input data. Part 1: Uncertainty analysis of the thermal-conductivity parameterization. Geothermics, 64, 42–54. https://doi.org/10.1016/j.geothermics.2016.04.010
48. Full, J., Hohmann, S., Ziehn, S., Gamero, E., Schließ, T., Schmid, H.-P., … Sauer, A. (2023). Perspectives of Biogas Plants as BECCS Facilities: A Comparative Analysis of Biomethane vs. Biohydrogen Production with Carbon Capture and Storage or Use (CCS/CCU). Energies, 16(13), 5066. https://doi.org/10.3390/en16135066
49. Fürhacker, M., Pressl, A., & Allabashi, R. (2003). Aerobic biodegradability of methyldiethanolamine (MDEA) used in natural gas sweetening plants in batch tests and continuous flow experiments. Chemosphere, 52(10), 1743–1748. https://doi.org/10.1016/S0045-6535(03)00371-0
50. Garcia, R., & Freire, F. (2014). Carbon footprint of particleboard: A comparison between ISO/TS 14067, GHG Protocol, PAS 2050 and Climate Declaration. Journal of Cleaner Production, 66, 199–209. https://doi.org/10.1016/j.jclepro.2013.11.073
51. Garrido-Barros, P., Gimbert-Suriñach, C., Matheu, R., Sala, X., & Llobet, A. (2017). How to make an efficient and robust molecular catalyst for water oxidation. Chemical Society Reviews, 46(20), 6088–6098. https://doi.org/10.1039/C7CS00248C
52. Gebauer, D., Völkel, A., & Cölfen, H. (2008). Stable Prenucleation Calcium Carbonate Clusters. Science, 322(5909), 1819–1822. https://doi.org/10.1126/science.1164271
53. González, E. J., Palomar, J., Navarro, P., Larriba, M., García, J., & Rodríguez, F. (2018). On the volatility of aromatic hydrocarbons in ionic liquids: Vapor-liquid equilibrium measurements and theoretical analysis. Journal of Molecular Liquids, 250, 9–18. https://doi.org/10.1016/j.molliq.2017.11.134

54. Grainger, A. (2015). Is Land Degradation Neutrality feasible in dry areas? Journal of Arid Environments, 112, 14–24. https://doi.org/10.1016/j.jaridenv.2014.05.014
55. Green, J. F. (2010). Private Standards in the Climate Regime: The Greenhouse Gas Protocol. Business and Politics, 12(3), 1–37. https://doi.org/10.2202/1469-3569.1318
56. Greschner, W., Lanzerath, C., Reß, T., Tenbrink, K., Borchert, S., Mix, A., … Gröger, H. (2014). Artificial cofactor regeneration with an iron(III)porphyrin as NADH-oxidase mimic in the enzymatic oxidation of l-glutamate to α-ketoglutarate. Journal of Molecular Catalysis B: Enzymatic, 103, 10–15. https://doi.org/10.1016/j.molcatb.2013.12.015
57. Gupta, J. (2010). A history of international climate change policy. WIREs Climate Change, 1(5), 636–653. https://doi.org/10.1002/wcc.67
58. Gupta, J., Olsthoorn, X., & Rotenberg, E. (2003). The role of scientific uncertainty in compliance with the Kyoto Protocol to the Climate Change Convention. Environmental Science & Policy, 6(6), 475–486. https://doi.org/10.1016/j.envsci.2003.09.001
59. Haas, P. M. (2000). International Institutions and Social Learning in the Management of Global Environmental Risks. Policy Studies Journal, 28(3), 558–575. https://doi.org/10.1111/j.1541-0072.2000.tb02048.x
60. Hanak, D. P., Biliyok, C., Yeung, H., & Białecki, R. (2014). Heat integration and exergy analysis for a supercritical high-ash coal-fired power plant integrated with a post-combustion carbon capture process. Fuel, 134, 126–139. https://doi.org/10.1016/j.fuel.2014.05.036
61. Hansen, J., Kharecha, P., Sato, M., Masson-Delmotte, V., Ackerman, F., Beerling, D. J., … Zachos, J. C. (2013). Assessing "Dangerous Climate Change": Required Reduction of Carbon Emissions to Protect Young People, Future Generations and Nature. PLOS ONE, 8(12), e81648. https://doi.org/10.1371/journal.pone.0081648
62. Harrould-Kolieb, E. R. (2019). (Re)Framing ocean acidification in the context of the United Nations Framework Convention on Climate Change (UNFCCC) and Paris Agreement. Climate Policy, 19(10), 1225–1238. https://doi.org/10.1080/14693062.2019.1649994
63. Hendriks, C. A., & Blok, K. (1992). Carbon dioxide recovery using a dual gas turbine IGCC plant. Energy Conversion and Management, 33(5), 387–396. https://doi.org/10.1016/0196-8904(92)90035-U

64. Herzog, H. J. (2011). Scaling up carbon dioxide capture and storage: From megatons to gigatons. Energy Economics, 33(4), 597–604. https://doi.org/10.1016/j.eneco.2010.11.004

65. Hickie, S. (2007, December 1). Maintaining Indigenous Peoples' Land Use and Management Rights in Kyoto Protocol Clean Development Mechanism Projects [SSRN Scholarly Paper]. Rochester, NY. https://doi.org/10.2139/ssrn.1351139

66. Hickmann, T. (2017). Voluntary global business initiatives and the international climate negotiations: A case study of the Greenhouse Gas Protocol. Journal of Cleaner Production, 169, 94–104. https://doi.org/10.1016/j.jclepro.2017.06.183

67. Höhne, N., Wartmann, S., Herold, A., & Freibauer, A. (2007). The rules for land use, land use change and forestry under the Kyoto Protocol—Lessons learned for the future climate negotiations. Environmental Science & Policy, 10(4), 353–369. https://doi.org/10.1016/j.envsci.2007.02.001

68. Holappa, L. (2020). A General Vision for Reduction of Energy Consumption and CO2 Emissions from the Steel Industry. Metals, 10(9), 1117. https://doi.org/10.3390/met10091117

69. Holtz-Eakin, D., & Selden, T. M. (1995). Stoking the fires? CO2 emissions and economic growth. Journal of Public Economics, 57(1), 85–101. https://doi.org/10.1016/0047-2727(94)01449-X

70. Howarth, R. W., Santoro, R., & Ingraffea, A. (2012). Venting and leaking of methane from shale gas development: Response to Cathles et al. Climatic Change, 113(2), 537–549. https://doi.org/10.1007/s10584-012-0401-0

71. Huang, Y., & Gerber, S. (2015). Global soil nitrous oxide emissions in a dynamic carbon-nitrogen model. Biogeosciences, 12(21), 6405–6427. https://doi.org/10.5194/bg-12-6405-2015

72. Hunt, A. J., Sin, E. H. K., Marriott, R., & Clark, J. H. (2010). Generation, Capture, and Utilization of Industrial Carbon Dioxide. ChemSusChem, 3(3), 306–322. https://doi.org/10.1002/cssc.200900169

73. Huq, S. (2002). The Bonn—Marrakech agreements on funding. Climate Policy, 2(2–3), 243–246. https://doi.org/10.3763/cpol.2002.0224

74. Idem, R., Wilson, M., Tontiwachwuthikul, P., Chakma, A., Veawab, A., Aroonwilas, A., & Gelowitz, D. (2006). Pilot Plant Studies of the CO2 Capture Performance of Aqueous MEA and Mixed MEA/

MDEA Solvents at the University of Regina CO2 Capture Technology Development Plant and the Boundary Dam CO2 Capture Demonstration Plant. Industrial & Engineering Chemistry Research, 45(8), 2414–2420. https://doi.org/10.1021/ie050569e

75. Jacobsson, T. J., Fjällström, V., Sahlberg, M., Edoff, M., & Edvinsson, T. (2013). A monolithic device for solar water splitting based on series interconnected thin film absorbers reaching over 10% solar-to-hydrogen efficiency. Energy & Environmental Science, 6(12), 3676–3683. https://doi.org/10.1039/C3EE42519C
76. Ješić, D., Lašič Jurković, D., Pohar, A., Suhadolnik, L., & Likozar, B. (2021). Engineering photocatalytic and photoelectrocatalytic CO2 reduction reactions: Mechanisms, intrinsic kinetics, mass transfer resistances, reactors and multi-scale modelling simulations. Chemical Engineering Journal, 407, 126799. https://doi.org/10.1016/j.cej.2020.126799
77. Jiang, G., Zhu, H., Zhang, X., Shen, B., Wu, L., Zhang, S., … Sun, S. (2015). Core/Shell Face-Centered Tetragonal FePd/Pd Nanoparticles as an Efficient Non-Pt Catalyst for the Oxygen Reduction Reaction. ACS Nano, 9(11), 11014–11022. https://doi.org/10.1021/acsnano.5b04361
78. Jiang, L., Liu, W., Wang, R. Q., Gonzalez-Diaz, A., Rojas-Michaga, M. F., Michailos, S., … Font-Palma, C. (2023). Sorption direct air capture with CO2 utilization. Progress in Energy and Combustion Science, 95, 101069. https://doi.org/10.1016/j.pecs.2022.101069
79. Kadier, A., Kalil, M. S., Abdeshahian, P., Chandrasekhar, K., Mohamed, A., Azman, N. F., … Hamid, A. A. (2016). Recent advances and emerging challenges in microbial electrolysis cells (MECs) for microbial production of hydrogen and value-added chemicals. Renewable and Sustainable Energy Reviews, 61, 501–525. https://doi.org/10.1016/j.rser.2016.04.017
80. Kaufman, B. E. (2003). High-level employee involvement at Delta Air Lines. Human Resource Management, 42(2), 175–190. https://doi.org/10.1002/hrm.10075
81. Keenan, T. F., & Williams, C. A. (2018). The Terrestrial Carbon Sink. Annual Review of Environment and Resources, 43(1), 219–243. https://doi.org/10.1146/annurev-environ-102017-030204
82. Keller, D. P., Lenton, A., Scott, V., Vaughan, N. E., Bauer, N., Ji, D., … Zickfeld, K. (2018). The Carbon Dioxide Removal Model Intercomparison Project (CDRMIP): Rationale and experimental

protocol for CMIP6. Geoscientific Model Development, 11(3), 1133–1160. https://doi.org/10.5194/gmd-11-1133-2018

83. Khaksar Manshad, A., Olad, M., Taghipour, S. A., Nowrouzi, I., & Mohammadi, A. H. (2016). Effects of water soluble ions on interfacial tension (IFT) between oil and brine in smart and carbonated smart water injection process in oil reservoirs. Journal of Molecular Liquids, 223, 987–993. https://doi.org/10.1016/j.molliq.2016.08.089
84. Kim, J. A. (2004). Regime interplay: The case of biodiversity and climate change. Global Environmental Change, 14(4), 315–324. https://doi.org/10.1016/j.gloenvcha.2004.04.001
85. Korppoo, A., & Gassan-Zade, O. (2014). Lessons from JI and GIS for post-2012 carbon finance mechanisms in Russia and Ukraine. Climate Policy, 14(2), 224–241. https://doi.org/10.1080/14693062.2014.844529
86. Kotir, J. H. (2011). Climate change and variability in Sub-Saharan Africa: A review of current and future trends and impacts on agriculture and food security. Environment, Development and Sustainability, 13(3), 587–605. https://doi.org/10.1007/s10668-010-9278-0
87. Lawn, P. (2016). International Climate Change Institutions and the Greenhouse-Gas Emitting Performance of Nations. In P. Lawn (Ed.), Resolving the Climate Change Crisis: The Ecological Economics of Climate Change (pp. 345–410). Dordrecht: Springer Netherlands. https://doi.org/10.1007/978-94-017-7502-1_8
88. Lee, K.-H. (2011). Integrating carbon footprint into supply chain management: The case of Hyundai Motor Company (HMC) in the automobile industry. Journal of Cleaner Production, 19(11), 1216–1223. https://doi.org/10.1016/j.jclepro.2011.03.010
89. Lenton, T. M., & Vaughan, N. E. (2009). The radiative forcing potential of different climate geoengineering options. Atmospheric Chemistry and Physics, 9(15), 5539–5561. https://doi.org/10.5194/acp-9-5539-2009
90. Letcher, T. M. (2020). 1—Introduction With a Focus on Atmospheric Carbon Dioxide and Climate Change. In T. M. Letcher (Ed.), Future Energy (Third Edition) (pp. 3–17). Elsevier. https://doi.org/10.1016/B978-0-08-102886-5.00001-3
91. Liang, Y., Liu, H., Rongwong, W., Liang, Z., Idem, R., & Tontiwachwuthikul, P. (2015). Solubility, absorption heat and

mass transfer studies of CO2 absorption into aqueous solution of 1-dimethylamino-2-propanol. Fuel, 144, 121–129. https://doi.org/10.1016/j.fuel.2014.11.098

92. Liao, Q., Liu, W., & Meng, Z. (2022). Strategies for overcoming the limitations of enzymatic carbon dioxide reduction. Biotechnology Advances, 60, 108024. https://doi.org/10.1016/j.biotechadv.2022.108024
93. Lim, B., Boileau, P., Bonduki, Y., van Amstel, A. R., Janssen, L. H. J. M., Olivier, J. G. J., & Kroeze, C. (1999). Improving the quality of national greenhouse gas inventories. Environmental Science & Policy, 2(3), 335–346. https://doi.org/10.1016/S1462-9011(99)00023-4
94. Littlejohns, T. J., Holliday, J., Gibson, L. M., Garratt, S., Oesingmann, N., Alfaro-Almagro, F., … Allen, N. E. (2020). The UK Biobank imaging enhancement of 100,000 participants: Rationale, data collection, management and future directions. Nature Communications, 11(1), 2624. https://doi.org/10.1038/s41467-020-15948-9
95. Liu, J., Bowman, K. W., Schimel, D. S., Parazoo, N. C., Jiang, Z., Lee, M., … Eldering, A. (2017). Contrasting carbon cycle responses of the tropical continents to the 2015–2016 El Niño. Science, 358(6360), eaam5690. https://doi.org/10.1126/science.aam5690
96. Locher, F., & Fressoz, J.-B. (2012). Modernity's Frail Climate: A Climate History of Environmental Reflexivity. Critical Inquiry, 38(3), 579–598. https://doi.org/10.1086/664552
97. Lu, S., Wang, Y., Xiang, H., Lei, H., Xu, B. B., Xing, L., … Liu, T. X. (2022). Mass transfer effect to electrochemical reduction of CO2: Electrode, electrocatalyst and electrolyte. Journal of Energy Storage, 52, 104764. https://doi.org/10.1016/j.est.2022.104764
98. Manley, B., & Maclaren, P. (2012). Potential impact of carbon trading on forest management in New Zealand. Forest Policy and Economics, 24, 35–40. https://doi.org/10.1016/j.forpol.2010.01.001
99. Marć, M., Kupka, T., Wieczorek, P. P., & Namieśnik, J. (2018). Computational modeling of molecularly imprinted polymers as a green approach to the development of novel analytical sorbents. TrAC Trends in Analytical Chemistry, 98, 64–78. https://doi.org/10.1016/j.trac.2017.10.020
100. Marković, N. M., Grgur, B. N., & Ross, P. N. (1997). Temperature-Dependent Hydrogen Electrochemistry on Platinum Low-Index

Single-Crystal Surfaces in Acid Solutions. The Journal of Physical Chemistry B, 101(27), 5405–5413. https://doi.org/10.1021/jp970930d

101. Massai, L. (2011). EC Compliance with the International Climate Regime. In L. Massai (Ed.), The Kyoto Protocol in the EU: European Community and Member States under International and European Law (pp. 137–239). The Hague, The Netherlands: T.M.C. Asser Press. https://doi.org/10.1007/978-90-6704-571-1_6
102. Matsuo, N. (2003). CDM in the Kyoto Negotiations: Mitigation and Adaptation Strategies for Global Change, 8(3), 191–200. https://doi.org/10.1023/B:MITI.0000005638.74001.45
103. Matuštík, J., & Kočí, V. (2021). What is a footprint? A conceptual analysis of environmental footprint indicators. Journal of Cleaner Production, 285, 124833. https://doi.org/10.1016/j.jclepro.2020.124833
104. McMillan, C., Boardman, R., McKellar, M., Sabharwall, P., Ruth, M., & Bragg-Sitton, S. (2016). Generation and Use of Thermal Energy in the U.S. Industrial Sector and Opportunities to Reduce its Carbon Emissions (No. NREL/TP-6A50-66763; INL/EXT-16-39680). National Renewable Energy Lab. (NREL), Golden, CO (United States). https://doi.org/10.2172/1335587
105. Metz, B., Berk, M., den Elzen, M., de Vries, B., & van Vuuren, D. (2002). Towards an equitable global climate change regime: Compatibility with Article 2 of the Climate Change Convention and the link with sustainable development. Climate Policy, 2(2–3), 211–230. https://doi.org/10.3763/cpol.2002.0221
106. Michael E. Parker, P. E., Northrop, S., Valencia, J. A., Foglesong, R. E., & Duncan, W. T. (2011). CO2 management at ExxonMobil's LaBarge field, Wyoming, USA. Energy Procedia, 4, 5455–5470. https://doi.org/10.1016/j.egypro.2011.02.531
107. Michaelowa, A. (2003). CDM host country institution building. Mitigation and Adaptation Strategies for Global Change, 8(3), 201–220. https://doi.org/10.1023/B:MITI.0000005639.89661.1c
108. Michaelowa, A., & Jotzo, F. (2005). Transaction costs, institutional rigidities and the size of the clean development mechanism. Energy Policy, 33(4), 511–523. https://doi.org/10.1016/j.enpol.2003.08.016
109. Milton, R. D., Wang, T., Knoche, K. L., & Minteer, S. D. (2016). Tailoring Biointerfaces for Electrocatalysis. Langmuir, 32(10), 2291–2301. https://doi.org/10.1021/acs.langmuir.5b04742

110. Neumann, B., Vafeidis, A. T., Zimmermann, J., & Nicholls, R. J. (2015). Future Coastal Population Growth and Exposure to Sea-Level Rise and Coastal Flooding—A Global Assessment. PLOS ONE, 10(3), e0118571. https://doi.org/10.1371/journal.pone.0118571

111. Oberthür, S. (2003). Institutional interaction to address greenhouse gas emissions from international transport: ICAO, IMO and the Kyoto Protocol. Climate Policy, 3(3), 191–205. https://doi.org/10.3763/cpol.2003.0327

112. Oka, R. (2021). Introducing an anthropology of convenience. Economic Anthropology, 8(2), 188–207. https://doi.org/10.1002/sea2.12219

113. Olajire, A. A. (2010). CO2 capture and separation technologies for end-of-pipe applications – A review. Energy, 35(6), 2610–2628. https://doi.org/10.1016/j.energy.2010.02.030

114. Ostrom, E. (2012). Nested externalities and polycentric institutions: Must we wait for global solutions to climate change before taking actions at other scales? Economic Theory, 49(2), 353–369. https://doi.org/10.1007/s00199-010-0558-6

115. Pant, T. S., Abhishek Sharma, Swati Sharma, Anand Giri, Ashok Kumar, Deepak. (2020). Recent Developments in CO2-Capture and Conversion Technologies. In Chemo-Biological Systems for CO2 Utilization. CRC Press.

116. Perrin, D., Temmerman, M., & Laitat, E. (2000). Calculation on the impacts of forestation, afforestation and reforestation on the C-sequestration potential in Belgian forests ecosystems. BASE. Retrieved from https://popups.uliege.be/1780-4507/index.php?id=15054

117. P. Gonçalves, L. C., R. Mansouri, H., PourMehdi, S., Abdellah, M., S. Fadiga, B., L. Bastos, E., … Rudroff, F. (2019). Boosting photobioredox catalysis by morpholine electron donors under aerobic conditions. Catalysis Science & Technology, 9(10), 2682–2688. https://doi.org/10.1039/C9CY00496C

118. Popp, D. (2003). Pollution control innovations and the Clean Air Act of 1990. Journal of Policy Analysis and Management, 22(4), 641–660. https://doi.org/10.1002/pam.10159

119. Portney, P. R. (1990). Air Pollution Policy. In Public Policies for Environmental Protection. Routledge.

120. Post, W. M., & Kwon, K. C. (2000). Soil carbon sequestration and

land-use change: Processes and potential. Global Change Biology, 6(3), 317–327. https://doi.org/10.1046/j.1365-2486.2000.00308.x

121. Priest, T. (2014). Hubbert's PeakThe Great Debate over the End of Oil. Historical Studies in the Natural Sciences, 44(1), 37–79. https://doi.org/10.1525/hsns.2014.44.1.37
122. Puxty, G., Rowland, R., & Attalla, M. (2010). Comparison of the rate of CO2 absorption into aqueous ammonia and monoethanolamine. Chemical Engineering Science, 65(2), 915–922. https://doi.org/10.1016/j.ces.2009.09.042
123. Qu, M., Abdelaziz, O., Sun, X.-G., & Yin, H. (2017). Aqueous solution of [EMIM][OAc]: Property formulations for use in air conditioning equipment design. Applied Thermal Engineering, 124, 271–278. https://doi.org/10.1016/j.applthermaleng.2017.05.167
124. Quadrelli, R., & Peterson, S. (2007). The energy–climate challenge: Recent trends in CO2 emissions from fuel combustion. Energy Policy, 35(11), 5938–5952. https://doi.org/10.1016/j.enpol.2007.07.001
125. Rani, B., Singh, U., Chuhan, A. K., Sharma, D., & Maheshwari, R. (2011). Photochemical Smog Pollution and Its Mitigation Measures. Journal of Advanced Scientific Research, 2(04), 28–33. Retrieved from https://sciensage.info/index.php/JASR/article/view/56
126. Rasouli, H., Nguyen, K., & Iliuta, M. C. (2022). Recent advancements in carbonic anhydrase immobilization and its implementation in CO2 capture technologies: A review. Separation and Purification Technology, 296, 121299. https://doi.org/10.1016/j.seppur.2022.121299
127. Rebitzer, G., Ekvall, T., Frischknecht, R., Hunkeler, D., Norris, G., Rydberg, T., … Pennington, D. W. (2004). Life cycle assessment: Part 1: Framework, goal and scope definition, inventory analysis, and applications. Environment International, 30(5), 701–720. https://doi.org/10.1016/j.envint.2003.11.005
128. Richter, B. D., Mathews, R., Harrison, D. L., & Wigington, R. (2003). Ecologically Sustainable Water Management: Managing River Flows for Ecological Integrity. Ecological Applications, 13(1), 206–224. https://doi.org/10.1890/1051-0761(2003)013[0206:ESWMMR]2.0.CO;2
129. Rickels, W., Proelß, A., Geden, O., Burhenne, J., & Fridahl, M. (2021). Integrating Carbon Dioxide Removal Into European Emissions Trading. Frontiers in Climate, 3. Retrieved from https://www.frontiersin.org/

articles/10.3389/fclim.2021.690023

130. Robine, J.-M., Cheung, S. L. K., Le Roy, S., Van Oyen, H., Griffiths, C., Michel, J.-P., & Herrmann, F. R. (2008). Death toll exceeded 70,000 in Europe during the summer of 2003. Comptes Rendus Biologies, 331(2), 171–178. https://doi.org/10.1016/j.crvi.2007.12.001
131. Romm, J. J. (2014). Cool Companies: How the Best Businesses Boost Profits and Productivity by Cutting Greenhouse Gas Emissions. London: Routledge. https://doi.org/10.4324/9781315071572
132. Rosen, A. M. (2015). The Wrong Solution at the Right Time: The Failure of the Kyoto Protocol on Climate Change. Politics & Policy, 43(1), 30–58. https://doi.org/10.1111/polp.12105
133. Rosswall, T. (2002, May 1). Greenhouse gases and global change: International collaboration. https://doi.org/10.1021/es00016a001
134. Rumberg, D. M., Bernhard Schlamadinger, Jenny L. P. Wong, Michael. (2005). Value and risks of expiring carbon credits from afforestation and reforestation projects under the CDM. In The EU Emissions Trading Scheme. Routledge.
135. Schaphoff, S., Lucht, W., Gerten, D., Sitch, S., Cramer, W., & Prentice, I. C. (2006). Terrestrial biosphere carbon storage under alternative climate projections. Climatic Change, 74(1), 97–122. https://doi.org/10.1007/s10584-005-9002-5
136. Schmalensee, R., Joskow, P. L., Ellerman, A. D., Montero, J. P., & Bailey, E. M. (1998). An Interim Evaluation of Sulfur Dioxide Emissions Trading. Journal of Economic Perspectives, 12(3), 53–68. https://doi.org/10.1257/jep.12.3.53
137. Schulze, E.-D., Valentini, R., & Sanz, M.-J. (2002). The long way from Kyoto to Marrakesh: Implications of the Kyoto Protocol negotiations for global ecology. Global Change Biology, 8(6), 505–518. https://doi.org/10.1046/j.1365-2486.2002.00523.x
138. Schwab, M. S., Gies, H., Freymond, C. V., Lupker, M., Haghipour, N., & Eglinton, T. I. (2022). Environmental and hydrologic controls on sediment and organic carbon export from a subalpine catchment: Insights from a time series. Biogeosciences, 19(23), 5591–5616. https://doi.org/10.5194/bg-19-5591-2022
139. Seo, S. N. (2017). Beyond the Paris Agreement: Climate change policy negotiations and future directions. Regional Science Policy & Practice, 9(2), 121–140. https://doi.org/10.1111/rsp3.12090

140. Sharkey, M., Harrad, S., Abou-Elwafa Abdallah, M., Drage, D. S., & Berresheim, H. (2020). Phasing-out of legacy brominated flame retardants: The UNEP Stockholm Convention and other legislative action worldwide. Environment International, 144, 106041. https://doi.org/10.1016/j.envint.2020.106041

141. Shiva Kumar, S., & Lim, H. (2022). An overview of water electrolysis technologies for green hydrogen production. Energy Reports, 8, 13793–13813. https://doi.org/10.1016/j.egyr.2022.10.127

142. Shorrock, L., & Dunster, J. (1997). The physically-based model BREHOMES and its use in deriving scenarios for the energy use and carbon dioxide emissions of the UK housing stock. Energy Policy, 25(12), 1027–1037. https://doi.org/10.1016/S0301-4215(97)00130-4

143. Smith, R. A. (2009). Saving the World or Corrupt Atmosphere Commodification? : The Legitimacy of Joint Implementation under Kyoto (Master thesis). Retrieved from https://www.duo.uio.no/handle/10852/22769

144. Spence, A., & Pidgeon, N. (2009). Psychology, Climate Change & Sustainable Behavior. Environment: Science and Policy for Sustainable Development, 51(6), 8–18. https://doi.org/10.1080/00139150903337217

145. Spooner, B., & Engel, D. (2011). Carbon Capture Using Amine-Based Technology. In Carbon Dioxide Sequestration and Related Technologies (pp. 121–131). John Wiley & Sons, Ltd. https://doi.org/10.1002/9781118175552.ch9

146. Sterk, W., & Schüle, R. (2009). Advancing the climate regime through linking domestic emission trading systems? Mitigation and Adaptation Strategies for Global Change, 14(5), 409–431. https://doi.org/10.1007/s11027-009-9178-5

147. Stott, P. A., Tett, S. F. B., Jones, G. S., Allen, M. R., Mitchell, J. F. B., & Jenkins, G. J. (2000). External Control of 20th Century Temperature by Natural and Anthropogenic Forcings. Science, 290(5499), 2133–2137. https://doi.org/10.1126/science.290.5499.2133

148. Stowell, D. (2005). Climate Trading. London: Palgrave Macmillan UK. https://doi.org/10.1057/9780230513846

149. Štreimikienė, D. (2013). The 18th session of the conference of the parties to the United Nations convention on climate change (UNFCCC). Intelektinė Ekonomika, 7(2), 254–259. Retrieved from https://www.

ceeol.com/search/article-detail?id=128051

150. Sun, C.-Y., & Chen, G.-J. (2005). Measurement of Interfacial Tension for the CO2 Injected Crude Oil + Reservoir Water System. Journal of Chemical & Engineering Data, 50(3), 936–938. https://doi.org/10.1021/je0495839
151. Swingland, I. R., Bettelheim, E. C., Grace, J., Prance, G. T., Saunders, L. S., Brown, S., … Myers, N. (2002). Changes in the use and management of forests for abating carbon emissions: Issues and challenges under the Kyoto Protocol. Philosophical Transactions of the Royal Society of London. Series A: Mathematical, Physical and Engineering Sciences, 360(1797), 1593–1605. https://doi.org/10.1098/rsta.2002.1021
152. Swingland, I. R., Bettelheim, E. C., Grace, J., Prance, G. T., Saunders, L. S., Malhi, Y., … Brown, S. (2002). Forests, carbon and global climate. Philosophical Transactions of the Royal Society of London. Series A: Mathematical, Physical and Engineering Sciences, 360(1797), 1567–1591. https://doi.org/10.1098/rsta.2002.1020
153. Tangen, K. (1999). The climate change negotiations: Buenos Aires and beyond. Global Environmental Change, 9(3), 175–178. https://doi.org/10.1016/S0959-3780(99)00008-4
154. T. Cousins, I., Goldenman, G., Herzke, D., Lohmann, R., Miller, M., A. Ng, C., … C. DeWitt, J. (2019). The concept of essential use for determining when uses of PFASs can be phased out. Environmental Science: Processes & Impacts, 21(11), 1803–1815. https://doi.org/10.1039/C9EM00163H
155. Thomas, K., Hardy, R. D., Lazrus, H., Mendez, M., Orlove, B., Rivera-Collazo, I., … Winthrop, R. (2019). Explaining differential vulnerability to climate change: A social science review. WIREs Climate Change, 10(2), e565. https://doi.org/10.1002/wcc.565
156. Thompson, J. M. T., & Malhi, Y. (2002). Carbon in the atmosphere and terrestrial biosphere in the 21st century. Philosophical Transactions of the Royal Society of London. Series A: Mathematical, Physical and Engineering Sciences, 360(1801), 2925–2945. https://doi.org/10.1098/rsta.2002.1098
157. Trenberth, K. E., & Fasullo, J. T. (2013). An apparent hiatus in global warming? Earth's Future, 1(1), 19–32. https://doi.org/10.1002/2013EF000165
158. Vaidya, P. D., & Kenig, E. Y. (2007). CO2-Alkanolamine Reaction

Kinetics: A Review of Recent Studies. Chemical Engineering & Technology, 30(11), 1467–1474. https://doi.org/10.1002/ceat.200700268

159. Vanek, F. M., Albright, L. D., & Angenent, L. T. (2016). Energy Systems Engineering: Evaluation and Implementation (3rd Edition). McGraw-Hill Education. Retrieved from https://www.accessengineeringlibrary.com/content/book/9781259585098
160. Vasileff, A., Zhi, X., Xu, C., Ge, L., Jiao, Y., Zheng, Y., & Qiao, S.-Z. (2019). Selectivity Control for Electrochemical CO2 Reduction by Charge Redistribution on the Surface of Copper Alloys. ACS Catalysis, 9(10), 9411–9417. https://doi.org/10.1021/acscatal.9b02312
161. Voigt, C. (2008, June 16). Is the Clean Development Mechanism Sustainable? Some Critical Aspects [SSRN Scholarly Paper]. Rochester, NY. Retrieved from https://papers.ssrn.com/abstract=1145262
162. von Gunten, U. (2003). Ozonation of drinking water: Part I. Oxidation kinetics and product formation. Water Research, 37(7), 1443–1467. https://doi.org/10.1016/S0043-1354(02)00457-8
163. Wall, T., Stanger, R., & Santos, S. (2011). Demonstrations of coal-fired oxy-fuel technology for carbon capture and storage and issues with commercial deployment. International Journal of Greenhouse Gas Control, 5, S5–S15. https://doi.org/10.1016/j.ijggc.2011.03.014
164. Watson, J. E. M., Dudley, N., Segan, D. B., & Hockings, M. (2014). The performance and potential of protected areas. Nature, 515(7525), 67–73. https://doi.org/10.1038/nature13947
165. Weart, S. R. (1997). The Discovery of the Risk of Global Warming. Physics Today, 50(1), 34–40. https://doi.org/10.1063/1.881664
166. Weir, G., & Muneer, T. (1998). Energy and environmental impact analysis of double-glazed windows. Energy Conversion and Management, 39(3), 243–256. https://doi.org/10.1016/S0196-8904(96)00191-4
167. Wettestad, J. (2009). Interaction between EU carbon trading and the international climate regime: Synergies and learning. International Environmental Agreements: Politics, Law and Economics, 9(4), 393–408. https://doi.org/10.1007/s10784-009-9107-9
168. Woerdman, E. (2001). Emissions trading and transaction costs: Analyzing the flaws in the discussion. Ecological Economics, 38(2), 293 304. https://doi.org/10.1016/S0921 8009(01)00169 0

169. Wright, L. A., Kemp, S., & Williams, I. (2011). 'Carbon footprinting': Towards a universally accepted definition. Carbon Management, 2(1), 61–72. https://doi.org/10.4155/cmt.10.39

170. Wuebbles, D. J., & Jain, A. K. (2001). Concerns about climate change and the role of fossil fuel use. Fuel Processing Technology, 71(1), 99–119. https://doi.org/10.1016/S0378-3820(01)00139-4

171. Wulf, S., Maeting, M., & Clemens, J. (2002). Application Technique and Slurry Co-Fermentation Effects on Ammonia, Nitrous Oxide, and Methane Emissions after Spreading. Journal of Environmental Quality, 31(6), 1795–1801. https://doi.org/10.2134/jeq2002.1795

172. Yamin, F. (2000). The role of the EU in climate negotiations. In J. Gupta & M. Grubb (Eds.), Climate Change and European Leadership: A Sustainable Role for Europe? (pp. 47–66). Dordrecht: Springer Netherlands. https://doi.org/10.1007/978-94-017-1049-7_4

173. Yang, H., Xu, Z., Fan, M., Gupta, R., Slimane, R. B., Bland, A. E., & Wright, I. (2008). Progress in carbon dioxide separation and capture: A review. Journal of Environmental Sciences, 20(1), 14–27. https://doi.org/10.1016/S1001-0742(08)60002-9

174. Youngman, R., Schmidt, J., Lee, J., & De Coninck, H. (2007). Evaluating technology transfer in the Clean Development Mechanism and Joint Implementation. Climate Policy, 7(6), 488–499. https://doi.org/10.1080/14693062.2007.9685672

175. Yuan, M., Kummer, M. J., & Minteer, S. D. (2019). Strategies for Bioelectrochemical CO2 Reduction. Chemistry – A European Journal, 25(63), 14258–14266. https://doi.org/10.1002/chem.201902880

176. Yun, H. J., Lee, H., Kim, N. D., Lee, D. M., Yu, S., & Yi, J. (2011). A Combination of Two Visible-Light Responsive Photocatalysts for Achieving the Z-Scheme in the Solid State. ACS Nano, 5(5), 4084–4090. https://doi.org/10.1021/nn2006738

177. Zhang, Z., & Baranzini, A. (2004). What do we know about carbon taxes? An inquiry into their impacts on competitiveness and distribution of income. Energy Policy, 32(4), 507–518. https://doi.org/10.1016/S0301-4215(03)00152-6

INDEX

B

C

D

E